T0341304

AQUATIC HEALTH AND AQUACULTURE

Healthy environments happen to cause better growth, quality production and higher yield. Fish and shellfish, being aquatic animals, their physiological activities are more or less directly controlled by many of the physico-chemical and biological properties of water and soil which in turn affect the growth and production of aquaculture. Therefore, this is the need of the hour to enhance aquatic productivity by sustainable intensification practices. Before adopting such practices in aquaculture industry, it is necessary to assess aquatic productivity and thus to help aquaculture operations especially in water quality monitoring and management and fish production. One of the major problems in modern aquaculture is outbreak of diseases and loss of crop due to poor water quality management. Maintaining conducive water quality and aquatic health is surely vital for successful and economically viable aquaculture operations. Such farming systems reduce the dependence on costly feeds and fertilizers and at the same time produce valuable fish and shellfish. The increased quantity and enhanced quality of fish produced by these practices can be a better option to withhold the nutritional security. The purpose to write this book is how to use the waters, wastewaters, liquid biowastes and soils unfit for agriculture to economically viable aquaculture practices; and putting the emphasis on, aquaculture posology, the science of quantification and administration of doses in aquatic health and aquaculture management. Broadly, aquaculture practices come across three types of problems each; in the context of water quality, and fin fish and shell fish diseases; floating, suspended and dissolved impurities in water; and preventive, curative and noncurative diseases in fin fish and shell fish. Keeping these views in the consideration, an attempt is made to write a book on, **Aquatic Health and Aquaculture**, which can be of immense help to the people working in this sector. Suggestions, improvements and amendments from users will be welcome.

Dr. Chandra Prakash, former Principal Scientist, ICAR-CIFE, Mumbai, was born on 4th March, 1956 at the city of Taj, Agra in Uttar Pradesh. He has the first class academic career and got awarded National Merit Scholarship, JRF, SRF and PDF of CSIR, New Delhi. He completed M.Sc. and Ph.D. from Agra University, Agra in the year 1976 and 1980 respectively. Having worked for a year period in Clear Water Ltd. Company, Delhi, joined Central Institute of Fisheries Education, Mumbai in the year 1983. He served in ICAR-CIFE for about 35 years and got retired as Principal Scientist in 2018. The main stream of his academic and research work was Aquatic Health, Environmental Science and Aquaculture. He guided the students for D.F.Sc., M.Sc., M.F.Sc., M.Phil. and Ph.D. programmes.

AQUATIC HEALTH AND AQUACULTURE

CHANDRA PRAKASH

ICAR-Central Institute of Fisheries Education
Panch Marg, Off Yari Road,
Versova, Andheri West
MUMBAI-400061, Maharashtra

CRC Press
Taylor & Francis Group
Boca Raton London New York

CRC Press is an imprint of the
Taylor & Francis Group, an **informa** business

NARENDRA PUBLISHING HOUSE
DELHI (INDIA)

First published 2021
by CRC Press
2 Park Square, Milton Park, Abingdon, Oxon, OX14 4RN
and by CRC Press
6000 Broken Sound Parkway NW, Suite 300, Boca Raton, FL 33487-2742

© 2020 Narendra Publishing House

CRC Press is an imprint of Informa UK Limited

The right of Chandra Prakash to be identified as author of this work has been asserted by him in accordance with sections 77 and 78 of the Copyright, Designs and Patents Act 1988.

Print edition not for sale in South Asia (India, Sri Lanka, Nepal, Bangladesh, Pakistan or Bhutan).

British Library Cataloguing-in-Publication Data
A catalogue record for this book is available from the British Library
Library of Congress Cataloging-in-Publication Data
A catalog record has been requested

ISBN: 978-0-367-62927-4 (hbk)
ISBN: 978-1-003-11152-8 (ebk)

NARENDRA PUBLISHING
HOUSE
DELHI (INDIA)

DEDICATED TO

Late Dr. S. N. DWIVEDI

Former Director, ICAR-CIFE, Mumbai, India

Contents

Foreword

Aquaculture is the fastest growing agriculture commodity with annual growth rate of more than 6% during last two decades. It is the most diverse sector in terms of species farmed, culture systems (freshwater, brackish water and marine), types and scales of operation, intensity of practices and types of management. Asia–Pacific region contributes about 89% of global aquaculture production. Tremendous progress has been achieved in aquaculture for the last three decades. This has been possible by increasing the area under farming and use of intensive and modern aquaculture practices involving higher use of inputs such as feeds, fertilizers and chemicals *etc*. However, stringent norms for satisfactory aquatic health through efficient water quality management practices are essential for sustainable growth of such ventures. Thus, water quality management is a vital factor in aquaculture in the context not only for intensive practices but also for burgeoning aquatic pollution problem and future scarcity of available water quantum.

The quantity of wastewaters being generated due to domestic and other anthropogenic activities coupled with salinization of soil and water sources is enormous. The wastewaters, being potentially polluted, may cause numerous negative impacts if released in to the environment directly. These wastewaters containing sewage; and dairy, sugar, beverage, confectionery and distillery industries' liquid biowastes, are rich in nutrients status and thus can be very well used for nutrients fed aquaculture which is essentially called Zero Feed-Zero Fertilizer system of farming. A considerable part of the wastes thus generated is liquid biowaste, which can be used for aquaculture in several ways. How to use, the water and wastewater scientifically successful for aquaculture; putting different quality control measures and techniques, is very well elaborated.

It is indeed a pleasure to see that Dr. Chandra Prakash, Principal Scientist (Retd.), ICAR-CIFE, Mumbai has written the book on, "**Aquatic Health and Aquaculture.**" The book prepared covers various aspects on aquatic health management like; water-soil/sediment analysis and interactions, integral calculation of doses, favourable ranges of various water quality parameters of all the three types of aquaculture practices and different water and wastewater quality control techniques in aquaculture and aquahatchery practices, which are the needs of the hour and justice to the title of the book. It will help to the persons engaged in teaching, research, extension and farming activities for aquaculture industry.

(A. G. Bhole)

Nagpur, India

13th, May, 2019

Former Professor of Environmental Engg. and HOD (civil) at VRCE, Nagpur and Emeritus Professor at LIT, Nagpur, India

Preface

Aquaculture is the fastest growing food producing industry in the world among the agriculture sectors with an annual growth rate of more than 6%. India is the second largest producer of aquaculture products like fish and shellfish production. Aquaculture was once considered to be environment friendly and socially responsible for food security. However, advent of modern technologies and intensification of the farming process have made aquaculture practices more competitive and modern innovations oriented. Intensification of culture practices in aquaculture needs the dissemination of aquatic health management activities through water quality management and control techniques to sustain its production. The importance of water quality and its control is two fold, first the intensive rearing and culture practices per unit of space and time; and secondly in the context of burgeoning problem of available water quality and quantity at the global level. As far as the quality of water is concerned, there are three types of impurities; the floating, suspended and dissolved. Removal of floating impurities needs screening; suspended particles require filtration and UV radiation (protozoan and microbial control) and dissolved impurities express necessity for chemical treatment, activated carbon and RO system.

Being aquatic animals, the physiological activities of fish and shellfish are more or less directly controlled by many of the physico-chemical and biological parameters of water and soil, which in turn affect the growth and production of aquaculture. Therefore, this is the need of the hour to enhance aquatic productivity by sustainable intensification practices. Before adopting such practices in aquaculture industry, it is necessary to assess its productivity and thus to help aquaculture operations especially in water quality management and fish production. One of the major problems in modern aquaculture is outbreak of diseases and loss of crop due to poor water quality management. Maintaining conducive water quality and aquatic health is vital for successful and economically viable aquaculture operations. Such farming systems reduce the dependence on costly feeds and fertilizers and at the same time produce valuable fish and shellfish. The increased quantity and enhanced quality of fish produced by these practices can be a better option to withhold the nutritional security. The purpose to write this book is how to use the waters, wastewaters, liquid biowastes and soils unfit for agriculture to economically viable aquaculture practices; and putting the emphasis on, aquaculture posology, the science of quantification and administration of doses in aquatic health and aquaculture management. Keeping these views in the consideration, an attempt is made to write a book on, **Aquatic Health and Aquaculture**, which can be of immense help to the people working in this sector. Suggestions, improvements and amendments from users will be welcome.

Chandra Prakash

CHAPTER 1

WATER DIVERSITY, CULTURE SYSTEMS AND PROBLEMS

"The concept of, Aquatic Health and Aquaculture, to enhance production per unit area/volume revolves around two basic principles of aquatic productivity, 1. Plankton turbidation of culture water and, 2. Benthos turbation of sediment."

Quantitative world water cycle

The total earth surface area is having almost ¾ part water and ¼ part land. In the land part, there are 26% mountains and hills, 33% plateau (table land) and the remaining 41% is plain land. From these figures, it appears that the earth is full of water and it can never face the problem of water scarcity. But the real picture is not like this. There are acute water scarcity problems for human consumption to have a look at the estimation of water availability. In the context of world water cycle, 5 lac km^3 of water gets vaporized from the earth surface. Out of 5 lac km^3, 4.3 lac km^3 is contributed by oceans and 0.7 lac km^3 by land part. When these vapours return to the earth surface after changing their state from clouds to rains, the ocean parts receive 3.9 lac km^3 and land part 1.1 lac km^3 of water. On these ways, the per year net water loss from the ocean surface is 0.4 lac km^3 and reciprocally there is net gain of the same amount of water on the land surface. Instead of getting so much additional quantity per year, land part receives only 22% of total rain water. Out of this, only a small quantity is available for human consumption. Thus, the water cycle gets involved three processes; vaporization, condensation and precipitation.

Classification of waters

Appearance-wise waters can be homogenous but composition-wise they are heterogeneous because water is the solution-mixture of various salts and universal solvent also, therefore, based on the nature, composition and quantum of dissolved and suspended solids available in waters; they can be classified as follows:

1. Freshwater, *i.e.*, such water in which concentration of dissolved solutes is lesser than 0.5 ppt/500 ppm/0.5 g or 500 mg per litre. Freshwater can be further subdivided into two categories.

    a)   Soft water in which carbonate and bicarbonate ions dominate.

    b)   Hard water in which sulphate ions dominate.

2. Saline water, *i.e.*, such water in which concentration of dissolved solutes is more than 0.5 ppt. Saline water can be further sub-divided into three categories.

    a)   Brackish water, salinity range is 0.5 ppt to 30 ppt.

    b)   Seawater, salinity is 30 ppt to 37 ppt.

    c)   Metahaline water or brine, salinity is >37 ppt.

3. Acidic water, which is dominated by the iron, aluminium and sulphur ions.

4. Wastewater or polluted water, the water unfit for human consumption.

    a)   Sewage, the liquid wastes discharged from domestic sources.

    b)   Effluent, the industrial wastewater.

5. Sludge, semi-liquid residue from industrial processes and treatment of water and   wastewater.

Table 1: Characteristics of raw acidic and hard water

Sl. No.	Parameter	Acidic water	Hard water
1.	Nature of water	Acidic	Alkaline
2.	pH	< 7.0	> 8.0
3.	Acidity (mg/l)	>250	< 50
4.	Alkalinity (mg/l)	< 50	> 275
5.	Total hardness (mg/l)	< 50	> 275
6.	Ionic dominance	Aluminium (Al^{+++}), Iron (Fe^{+++}) and Sulphur (S^{--})	Calcium (Ca^{++}) and Magnesium (Mg^{++})

Types of culture systems

Aquaculture is practised in all the three types of aquatic environments.

1. **Freshwater aquaculture:** Freshwater, *i.e.*, waters having salinity level lesser than 0.5 parts per thousand (ppt).

2. **Brackish water aquaculture:** Brackish water with salinity range of 0.5 to 30 ppt.

3. **Mariculture** or **sea farming:** Seawater with more than 30 ppt salinity.

The species of flora and fauna inhabiting these three types of water bodies are accordingly called freshwater species, brackish water species and marine species. Freshwater, which is the most extensively used sector of aquaculture, is further divided into two segments.

a) **Cold waters (temperate aquaculture)** of higher altitudes having temperature range lesser than 18^0C and,

b) **Warm waters (tropical aquaculture)** of plains having temperature range of more than 18^0C.

Aquaculture practices in these waters are, therefore, called coldwater aquaculture and warm water aquaculture respectively. Freshwater aquaculture is carried out in fishponds, fish pens, fish cages and on a limited scale in paddy fields. Brackish water aquaculture is done mainly in fish ponds located in coastal areas. Marine culture employs either fish cages or substrates for molluscs and seaweeds such as stakes, ropes and rafts.

Aquaculture is practised through various methods. Culture of fishes in ponds is the oldest form of aquaculture. There are three ways in which pond aquaculture is conducted.

i)   Aquaculture in un-drainable ponds

ii) Aquaculture in drainable ponds and

iii) Aquaculture having recirculating water.

Aquaculture in pens is a method in which aquatic organisms are grown in specially erected enclosure in open water bodies. Aquaculture in cages, likewise, is a method in which cultivable organisms are raised in enclosures installed in suspended state in flowing or stagnant waters. Free exchange of water is maintained in pens and cages both. Integrated aquaculture is an innovative approach in which aquatic organisms are grown in combination with agriculture or livestock, raising in such a way that the by-products of each system mutually benefit the other. Molluscs such as oysters and mussels are cultured on suspended ropes or wires hung from rafts floating at the surface of water. The seed, *i.e.,* bivalve larvae attached on shells are strung on ropes and hung in water for further growth. Additional floats are added to the raft as the growth proceeds and total weight of the growing biomass increases. Seaweeds are grown using different types of planting material such as vegetative cuttings, natural seed and hatchery-reared seed. The methods of culture include bottom culture, rope culture and pond culture either in monoculture or polyculture with milkfish, shrimps and crabs.

Different levels of aquaculture

Depending on the intensity of operation and degree of management, aquaculture practices are classified into three operation scales *viz.*

i)   Extensive Aquaculture

ii) Semi-intensive Aquaculture

iii) Intensive Aquaculture.

Extensive level

In extensive level of aquaculture, low stocking densities of 2000-5000 carp fry/ ha or 5000 to 10,000 shrimps post larvae (PL/ha/crop) are used and no supplemental feed is given. Fertilization (application of fertilizers) may be done to stimulate the growth and production of natural food in the water. Unlike shrimp culture, carp culture may not require water exchange during culture period. In shrimp culture, water exchange is done through tidal flushing, *i.e.,* new water is allowed to let in during high tide and the used water is drained out during the low tide. The ponds used for extensive aquaculture are usually large (more than two hectare). The production is generally low, lesser than 0.5 ton/ha/year in the case of shrimps and 1 to 3 ton/ha/year in the case of carp.

Semi-intensive level

Semi-intensive aquaculture is practised in medium sized ponds of 0.5 ha each with comparatively higher stocking densities (*i.e.,* 50,000 to 1, 00,000 shrimp PL/ ha or 10,000 carp fingerlings/ha) than extensive aquaculture. Supplementary feeding is done in moderate amount. In the case of carp culture, water exchange is done once or twice a month @ 10%, while in shrimp culture; it is done once or twice every week @ 10 to 20%. The production is around 1.5 ton/ha in shrimp culture and 5 to 7 ton/ha in the case of carps.

Intensive level

In intensive level of aquaculture, the pond size for the case of operational convenience is generally small, about 0.2 ha approximately, with very high density of culture organisms, *i.e.*, 300,000 to 500,000 shrimp PL/ha or 20,000 to 25,000 carp fingerlings/ha. The system is totally dependent on the use of formulated feeds. Feeding of the stock is done at regular intervals. In intensive shrimp culture, the computed daily feed ration is given in equal amounts from as low as three to as high as six times a day. Water replacement under intensive culture is done on a daily basis approximately @ 25 to 30%. Production under intensive level of aquaculture is much higher, for example, 8 to 10 ton/ha/crop in shrimp culture and about 12 to 15 ton/ha in carp culture.

Super intensive level

Super intensive aquaculture needs running water supply and complete daily water exchange is performed. This system is mostly practiced in cement tanks, fiberglass tanks and raceways *etc.*, which are fitted with high efficiency biological filters for continuous recirculation of water. The size of the tank ranges between 50-100 m^3. The cultured organisms are fed with high quality formulated feed, which is given through demand feeders. The water quality is regularly monitored with electronic gadgets. Stocking density ranges between 8, 00,000 to 10, 00,000 shrimp PL/ha or 40,000 to 50,000 carp fingerlings/ha. The production ranges between 10 to 12 ton/ha in shrimp culture and 15 to 20 ton/ha in case of carps.

Water quality problems and aquaculture

The common water quality problems in the context of aquaculture practices are due to the nonoptimal levels of following given parameters; 1. Colour, 2. Clay turbidity, 3. pH, 4. Alkalinity, 5. Chloride, 6. Hardness, 7. Ammonia, 8. Nitrite, 9. Carbondioxide, 10. BOD, 11. COD, 12. Iron, 13. Residual chlorine and 14. Organic carbon in sediments. Their effects and methods of removal are discussed in the different chapters ahead.

FAO has defined aquaculture as, the farming of aquatic organisms, including fish, molluscs, crustaceans and aquatic plants. Farming implies some form of intervention in the rearing process to enhance production, such as regular stocking, feeding and protection from predators *etc*. Broadly, aquaculture practices come across three types of problems each; in the context of water quality, fin fish and shell fish diseases. Floating, suspended and dissolved impurities in water; and preventive, curative and noncurative diseases in fin fish and shell fish. In this book, efforts are put to frame the solutions of these problems.

CHAPTER 2

IMPORTANT UNITS AND DIMENSIONS

Following are the important units and dimensions commonly used in aquatic environment health and water quality management practices. One litre of pure water at 4°C temperature and normal atmospheric pressure weighs exactly 1 Kg.

Mass

ppt - parts per thousand or gm/l

ppm - parts per million or mg/l

1 pound = 0.4535 kg = 453.5 gm

1 pound = 16 oz. (ounce)

1 oz. (ounce) = 28.3495 gm

1 ppm = μg/g = mg/kg = g/ton = mg/litre = 1 ml/1000 liters (1 ml/m^3) =1000 ppb

1 gm = 1000 mg

1 mg = 1000 micro gram (μg)

1 ppb = 1 μg/l or 1 mg/m^3

1 percent = 10,000 ppm

Length/particle size

1 Fathom = 6 feet

1 Km = 0.6214 mile = 0.540 nautical mile

1 nautical mile = 1.85 km

1 Knot = 51.5 cm/sec = 1 nautical mile/hour

1 m = 3.28 feet

1 cm = 10 mm = 0.394 inch

1 inch = 25.4 mm = 2.54 cm

1 foot = 12 inch = 30.48 cm

3 feet = 1 yard

1 yard = 91.44 cm = 0.9144 meter

1 mm = 1000 micron

1 micron = 10^{-6} m

Micro = 10^{-6} (prefix)

1 angstrom = 0.0001 micron = 0.1 nanometer = 1 x 10^{-10} meter

Area

1 hectare = 10,000 m^2

1 hectare = 2.47 acres

1 square mile = 2.590 sq Km

1 square meter = 1555 sq inch = 10.764 sq feet

1 square foot = 929.0304 sq cm

Rectangle = Length x width

Cylinder = $2\pi rh$

Sphere = $4\pi r^2$

Trapezium = 1/2 sum of parallel sides x height

Volume

1 m^3 = 1000 liters

1 m^3 = 35.307 ft^3

1 ft^3 = 28.3168 liters

1 litre = 1000 ml or CC

1 CC or ml = 1000 cubic millimeter

1 gallon = 4.546 liters (British Unit)

1 gallon = 3.785 liters (U.S. Unit)

1 cusec = 1 ft^3/sec.

1cumec = 1 m^3/sec.

1 lpm = 1 litre per minute

Dissolved oxygen 1 ml = 1.43 mg*

Rectangle = Length x width x height

Cylinder = πr^2 h

Sphere = $4/3\pi r^3$

Trapezium = 1/2 sum of parallel sides x height x length

* The mass of one litre oxygen at normal atmospheric pressure and temperature is 1.43 g,

Thus, 1 litre oxygen = 1.43 g,

1000 ml = 1430 mg,

1 ml = 1.43 mg.

Table 2: Application of metric units

Description	Unit	Symbol
Precipitation run-off	Milli meter (1mm of rain = 1liter/m^2)	mm
River flow	Cubic meter per second	M^3/s
Flow in pipes	Cubic meter per second or liter per second	M^3/s or L/s
Usage of water	Litre per person per day	L/person/day
Density	Kilogram per cubic meter	Kg/M^3 or g/ml
Concentration	Milligram per litre or gram per cubic meter	mg/L or g/M^3
B.O.D. loading	Kilogram per cubic meter per day	Kg/M^3/d
Hydraulic load	Cubic meter per square-meter per day	M^3/M^2/d
Per unit area hydraulic load	Cubic meter per square-meter per day	M^3/M^2/d
Per unit volume velocity	Meter per second	m/s

CHAPTER 3

MONITORING OF WATER AND SEDIMENT PARAMETERS

INTRODUCTION

The analysis of water and wastewater plays a significant role in aquatic environment and aquaculture studies. It helps to diagnose the constraints in the aquatic system and to explore management options for optimum economic returns. At this stage of starting the analytical work, the points to be kept in consideration are laboratory safety, proper sampling, labeling and appropriate storage of the samples collected and the requisite chemicals and solutions. To prepare the solutions of known strength, the normality, atomic weight, molecular weight, equivalent (combining) weight and valency of common elements, radicals and compounds should be known. The importance of water quality control is twofold, first in the context of burgeoning aquatic pollution problem and secondly with reference to intensive rearing and culture per unit area/volume of water. Good water quality provides better growth and higher survival rate to cultivable organisms. The water quality management is a three tier system comprising the analysis of various water parameters, to have the knowledge of the favourable ranges of these parameters and thirdly the chemical treatment and control techniques involved to bring the parameters under desirable limits.

The collected sample analysis is the partial representation of precise available facts, not the real and comprehensive depiction of true facts, present in source water. From the observed and analysed data, the inferences are to be drawn to implement. Physico-chemical and biological properties of an aquatic ecosystem play a significant role in its productivity process and in turn the growth of aquatic organisms under culture. Selection of culture for aquaculture species is basically made on the basis of prevailing physico-chemical properties of the environment of a particular aquatic ecosystem. These parameters vary on the basis of geographical location, season and seasonal changes and the nature of individual water bodies. The prime difference between freshwater and saline water is in their salt contents and composition of ionic concentration.

Objective

To estimate the important water quality parameters like temperature, turbidity, transparency, light, depth, pH, dissolved oxygen, dissolved free carbon dioxide, biochemical oxygen demand (BOD), chemical oxygen demand (COD), total dissolved solids, salinity, chlorides, alkalinity, hardness, ammonia-N, nitrite-N, nitrate-N, phosphorus in the given water sample; and texture, pH and organic carbon in soil/sediment.

PROCEDURE FOR ESTIMATION

Temperature

It plays a very important role in aquatic studies and usually determined by a centigrade thermometer having a least count of 0.1^0C. Surface water temperature can be observed immediately after collection and dipping the thermometer directly into collected sample water. Subsurface water temperature can be known by a reverse thermometer. Temperature affects the chemical changes in sediment and water including composition contents and pressure of dissolved gases. Selection of cultivable species depends on its body temperature requirement or tolerance limit, which is ± 1 or 2^0C of ambient temperature. The optimal temperature range for temperate and tropical fishes is about 10 to 18^0C and 25 to 32^0C respectively. The optimal temperature requirement for shrimp in coastal aquaculture is about 28 to 32^0C.

Turbidity, transparency and light

These first two physical properties of water are inversely proportional to each other. Field level turbidity can be known by turbidity meter rod while at laboratory level by nephalometer. Turbidity meter rod value is expressed in ppm unit; on the other hand, it is NTU (Nephelometric Turbidity Unit) for nephalometeric estimation. One NTU is about 2.5 to 3.0 ppm of turbidity meter rod value. The U.S. Geological Survey turbidity rod is commonly used to measure turbidity at field level. It is a 6 mm thick, 17 mm wide and 1 meter long metal rod marked with 10 to 3000 ppm. A piece of straight, bright platinum wire of 1 mm diameter and of 25 mm length is attached in a detachable screw at the lower end of scale. To view this platinum wire an eye point is marked at the upper end of scale. To measure the turbidity, hold the rod straight and dip it in water vertically; now observe the platinum wire from eye point mark. When the visibility of platinum wire fades, the water level on the graduated turbidity rod is recorded. Thus, the value obtained is the turbidity measurement of sample water in ppm.

Transparency is estimated by Secchi disc and expressed in the unit, cm. The Secchi disc, as created by Angelo Secchi in the year 1865, is a plain white circular disc of 30 cm (12 inch) in diameter which is used to measure water transparency in aquatic bodies. This measure is known as the Secchi depth/ transparency, the clarity of water and it is related to light penetration in water bodies. Since its invention, the disc has also been used in a modified, smaller 20 cm (8 inch) diameter black and white plate designed to measure water transparency. George C. Whipple in the year 1899 modified the original all-white Secchi disc to a disc of 20 cm (8 inch) in diameter, divided into four quadrants painted alternately black and white. Modified black and white Secchi disc is the standard disc now used in hydrobiological studies. A staple is fixed in the centre at the upper surface of disc which provides attachment to graduated rope. Opposite to staple, on the lower side of disc, sinker weight (anti buoyant) is fixed that facilitates sinking of disc in water while operating it. To measure transparency, dip the disc into water by holding graduated rope. Note the water level on graduated rope when the disc gets disappeared. Now lift the disc slowly and note the water level again on graduated rope when the disc gets reappeared. The average of these two values is the transparency in cm. Transparency (cm) = Disappearance value of disc (cm) + Reappearance value of disc (cm)/2. Light intensity and photoperiod, both are equally important for vegetative and animal growth. For better aquatic primary productivity, an illumination of 5000 to 20,000 lux is in suitable range. The light intensity can be observed by lux meter.

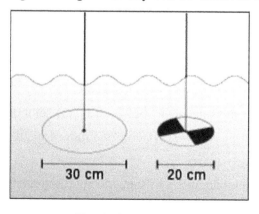

Fig. 1: Secchi disc

Depth

It has a significant impact on water properties of aquatic bodies and thus plays a major role in aquaculture. Its requirement differs from organism to organism under culture practices. Carps require about 1.5 meter depth, while for shrimp 1 meter depth is adequate. Air breathing fishes need shallow waters in order to perform surfacing action for inhaling atmospheric oxygen.

pH

It is a logarithmic scale used to specify acidity or basicity of an aqueous solution and the negative of base 10 logarithm of the molar concentration, measured in unit of moles per litre of hydrogen ions; $pH = -log_{10}[H^+]$. In literary terms, it is called as pulse rate of water quality. The pH determination can be done by electrometric and colorimetric methods. Although it expresses the intensity of acid or base reaction of a solution in terms of hydrogen ion concentration yet it is not a measure of total concentration of acidity or alkalinity. Electrometer can determine comparatively more accurate results for clear as well as coloured and turbid waters than colorimetric estimation. The interference created due to colour and turbidity in the water and wastewater samples for pH estimation can be brought down by applying a pinch of neutral barium sulphate ($BaSO_4$); thus get the testing sample as clear liquid after stirring and then allow to settle for 10 to 15 minutes. After decantation pH value can be tested.

ESTIMATION OF DISSOLVED OXYGEN

Material required

i. Manganous sulphate (Winkler's – A) solution.

    Dissolve 480 g of $MnSO_4.4H_2O$ or 400 g of $MnSO_4.2H_2O$ or 364 g of $MnSO_4.H_2O$ in one litre of distilled water.

ii. Alkaline potassium iodide (Winkler's – B) solution.

    Dissolve 500 g of NaOH, 150 g of KI (Potassium iodide) and 20 g of NaN_3 (Sodium azide) in one litre of distilled water.

iii. Standard sodium thiosulphate (0.025 N or N/40) solution.

    Dissolve 6.205 g of $Na_2S_2O_3.5H_2O$ in one litre of distilled water.

iv. Starch indicator.

    Make a thin paste of about 1 g of starch in cold water, pour 200 ml of boiling water in it and stir, add 0.1 g of salicylic acid or 0.5 g of sodium chloride or 0.5 ml of formalin as a preservative.

v. Burette, 1 ml and 10 ml; pipette, 250 ml; conical flask; DO bottles and a white tile.

N.B.: 1 ml/litre DO = 1.43 mg/lit

Procedure

Remove carefully the stopper of the 100 ml sample bottle (containing water sample for O_2), add 1 ml of manganous sulphate reagent and 1 ml of alkaline iodide reagent by means of one ml pipette dipped to the bottom of the bottle and slowly drawing out as the reagents are added. Replace the stopper and invert the bottle, three or four times for a thorough mixing of the reagents. A flocculant precipitate will be formed which will settle at the bottom. If the precipitate is whitish in colour, oxygen is very poor, light brown colour indicates high dissolved oxygen. For quantitative estimation, add 1 ml of concentrate H_2SO_4 to dissolve the precipitate. Transfer 50 ml of the solution to a conical flask placed on a white background (white tile), add N/40 sodium thiosulphate solution drop by drop till the colour turns pale yellow. Then add 1 ml of starch indicator solution to give a blue colour and titrate it till it is colourless. Note down the number of ml of $Na_2S_2O_3.5H_2O$ solution consumed.

Calculation

$$\text{Dissolved oxygen} = \frac{\text{No. of ml of } Na_2S_2O_3.5H_2O \text{ solution consumed} \times 1000 \times 8}{50 \times 40} \text{mg/l}$$

Dissolved oxygen = No. of ml of $Na_2S_2O_3.5H_2O$ solution consumed $\times$ 4 mg/l

ESTIMATION OF DISSOLVED FREE CARBON DIOXIDE

Material required

i.   Phenolphthalein indicator (dissolve 5 g of phenolphthalein in one litre of 50% alcohol (C_2H_5OH) or 0.5 g in 100 ml).

ii. N/44 NaOH, sodium hydroxide (dissolve 0.909 g of NaOH pellets in one litre of distilled water).

iii. 1 ml pipette, conical flask and a white tile.

Procedure

Take 50 ml of the sample in a Nessler's tube and add two drops of phenolphthalein indicator. If the water turns pink, there is no free carbon dioxide. If the water remains colourless, add $^N/_{44}$ NaOH drop by drop through a 1 ml graduated pipette with a very gentle stirring using a glass rod till the colour turns pink. Record ml of $^N/_{44}$ NaOH used during this titration.

Calculation

$$\text{Dissolved free carbon dioxide} = \frac{\text{No. of ml of N/44 NaOH used} \times 1000 \times 44}{50 \times 44} \text{mg/l}$$

Dissolved free carbon dioxide = No. of ml of N/44 NaOH used x 20 mg/l

ESTIMATION OF BIOCHEMICAL OXYGEN DEMAND (BOD)

Material required

i.   Phosphate buffer (dissolve 8.5 g of KH_2PO_4, 21.75 g of K_2HPO_4 and 33.4 g of $Na_2HPO_4.7H_2O$ in 500 ml of distilled water and dilute it to one litre).

ii.   Magnesium sulphate solution (dissolve 22.5 g of $MgSO_4.7H_2O$ in subtle distilled water and dilute to one litre).

iii. Calcium chloride solution (dissolve 27.5 g of anhydrous calcium chloride in subtle distilled water and dilute to one litre).

iv.   Ferric chloride solution (dissolve 0.25 g $FeCl_3.6H_2O$ in sufficient distilled water and dilute to one litre).

Procedure

i.   Prepare dilution water by adding 1.0 ml each of the solutions, phosphate buffer, magnesium sulphate, calcium chloride and ferric chloride to one litre of distilled water.

ii.   Add 2.0 ml of sewage and aerate. Determine the exact capacity of three 125 ml narrow mouthed flat stopper bottles (NMFS).

iii. Find out DO of undiluted sample.

iv.   Prepare the desired mixture by adding sample in dilution water.

v.   Fill up one bottle with the mixture and the other one with dilution water (blank).

vi.   Incubate at any protocol (20°C, 5 days; 35°C, 3days; 27°C, 3days).

vii. Find out DO in both the bottles after incubation.

Table 3:   BOD measurable with various dilutions of samples using percent mixture/
direct pipetting

% mixture	Range of BOD (mg/l)	ml	Range of BOD (mg/l)
0.01	20,000-70,000	0.02	30,000-105,000
0.02	10,000-35,000	0.05	12,000-42,000
0.05	4,000-14,000	0.10	6,000-21,000
0.1	2,000-7,000	0.2	3,000-10,500
0.2	1,000-3,500	0.5	1,200 -4,200
0.5	400-1,400	1.0	600-2,100
1.0	200-700	2.0	300-1,050
2.0	100-350	5.0	120-420
5.0	40-140	10	60-210
10	20-70	20	30-105
20	10-35	50	12-42
50	4-14	100	6-21
100	0-7	300	0-7

*20^0C – 5 days BOD is about 69% of the total demand.

*20^0C -10 days BOD is almost 90 % estimation of the total demand.

*As per international norms BOD load per person per day is 0.1 kg.

*As per the gazette notification of the Ministry of Environment and Forest, Govt. of India, 1998, BOD can be observed at 27^0C temperature for 3 days period.

Calculation

$$BOD \text{ in mg/l} = \frac{((Dob-Doi)\ 100) - Dob - D0s)}{\%} \text{ (For percent mixture)}$$

$$BOD \text{ in mg/l} = \frac{((Dob-Doi)\ Vol.\ of\ bottle - Dob - D0s)}{Vol.\ of\ sample} \text{ (For direct pipetting)}$$

Where: Dob – DO* of blank (dilution water) after incubation.

Doi – DO of mixture after incubation.

Dos – DO of undiluted sample.

---

*DO: dissolved oxygen

ESTIMATION OF CHEMICAL OXYGEN DEMAND (COD)

Material required

i.   Potassium permanganate solution, $KMnO_4$ (0.1N), dissolve 2.9 g in 1litre distilled water.

ii. Sulphuric acid, H_2SO_4 (25% v/v).

iii. Potassium iodide, KI (10%).

iv. Sodium thiosulphate solution, $Na_2S_2O_3.5H_2O$ (0.1 N), dissolve 24.8 g in 1litre distilled water.

Procedure

Take 50 ml of water sample in a 250 ml conical flask and 50 ml of distilled water in another flask to run the blank. Add 5 ml of $KMnO_4$ (0.1 N) solution to both flasks and heat them on a water bath at boiling point for one hour and then cool to room temperature. Add 5 ml of potassium iodide (10%) solution in both the flasks followed by 10 ml of (25% v/v) sulphuric acid. Titrate both, sample and the blank with 0.1 N sodium thiosulphate solution using starch as indicator. The blue colour produced by starch would disappear sharply at the end point. Note the reading for samples (S) and Blank (B). Repeat the observations 3 times for concordant results. Calculate the COD as mentioned below:

Calculation

$$COD \text{ in mg/l} = \frac{(B\text{-}S) \times N \times 8 \times 100}{\text{Sample volume in ml}}$$

Where: B is the volume of titrant used in blank.

S is the volume of titrant used for sample.

N is strength of the titrant.

ESTIMATION OF SOLIDS

Material required

Porcelain dish, silica crucible, corning or pyrex beaker (500 ml capacity), funnel, oven with thermostatic control, desiccator, Whatman filter paper No.44, physical and analytical balance *etc.*

Procedure

For total solids (TS)

i.   Place the sample in a weighed dry dish/crucible/beaker and evaporate to dryness in an oven at 103°C to 105°C.

ii. Cool the container to dryness in a desiccator.

iii. Weigh the dish/crucible/beaker and note the increase in weight.

$$\text{Total solids} = \frac{\text{mg residue x 1000}}{\text{ml of sample}} \text{ mg/l}$$

For total dissolved solids (TDS)

i.   Filter the sample water through Whatman filter paper No.44.

ii. Take a suitable quantity in a weighed dry dish/crucible/beaker (Weight No.1) and evaporate to dryness in an oven at 103°C to 105°C.

iii. Cool the container to dryness in a desiccator.

iv. Take the weight of dried dish/crucible/beaker (Weight No. 2).

v. $$\text{Total dissolved solids} = \frac{\text{Weight No.2} - \text{Weight No.1} \times 1000}{\text{ml of sample}} \text{ mg/l}$$

For total suspended solids (TSS)

Total suspended solids = total solids − total dissolved solids mg/l

ESTIMATION OF SALINITY

Material required

i.   Silver nitrate solution of 0.28N (dissolve 48.5 g of $AgNO_3$ in one litre of distilled water).

ii. Potassium chromate indicator (dissolve 63 g of K_2CrO_4 in 100 ml of distilled water).

Procedure

Pipette out 10 ml of sample water in a flask and add 2 or 3 drops of K_2CrO_4 indicator; colour of sample turns yellow. Titrate it against 0.28 N $AgNO_3$ drop by drop so that the solution becomes reddish brown which does not disappear on shaking. This is the end point.

$$AgNO_3 + NaCl = AgCl + NaNO_3$$

$$2AgNO_3 + K_2CrO_4 = Ag_2CrO_4 + 2KNO_3$$

Calculation

Salinity (ppt) = Burette reading + 0.0355 (correction factor)

ESTIMATION OF CHLORIDES

Material required

i.   Silver nitrate solution of N/35 (dissolve 4.78 g of $AgNO_3$ in one litre of distilled water).

ii. Potassium chromate indicator (dissolve 50 g of K_2CrO_4 in one litre of distilled water).

Procedure

Place 100 ml or more of sample in two conical flasks each. Add two drops of potassium chromate indicator to each of the flasks. Titrate against the liquid in one flask with N/35 $AgNO_3$ solution from a burette drop by drop with constant stirring until a permanent change in colour is produced (by comparing it with the second flask). The colour will get changed from yellow to brick red. Record the ml of $AgNO_3$ solution used.

Calculation

$$\text{Chlorides (Cl}^-) = \frac{AgNO_3 \text{ solution consumed x 1000}}{\text{ml of sample}} \text{ mg/l}$$

ESTIMATION OF ALKALINITY

Material required

i.   Sulphuric acid (N/50) – First, prepare N H_2SO_4 (stock solution) by adding one ml of concentrate H_2SO_4 in 35 ml distilled water. Now take 10 ml of this N H_2SO_4 (stock solution) and dilute to 500 ml with distilled water.

ii. Phenolphthalein indicator (dissolve 5 g of phenolphthalein in one litre of 50% alcohol (C_2H_5OH) or 0.5 g in 100 ml).

iii. Methyl orange indicator (dissolve 5 g of methyl orange in one litre of distilled water).

Procedure

Place 100 ml of sample water each in two conical flasks. In one flask, add 0.5 ml phenolphthalein indicator. If the sample becomes pink, titrate it with $N/_{50}$ H_2SO_4 from burette until the pink colour just disappears. Record the ml of acid used. Add two drops of methyl orange indicator in both the conical flasks and titrate one of them with $N/_{50}$ H_2SO_4. The end point is orange. The end point can be best judged by comparing it with blank. Record the ml of acid used both in phenolphthalein and methyl orange titration.

Calculation

$$\text{Total alkalinity} = \frac{\text{No. of ml of N/50 } H_2SO_4 \text{ consumed x 1000}}{\text{ml of sample}} \text{ mg/l}$$

ESTIMATION OF HARDNESS

Material required

i. Erichrome black-T indicator, dissolve 0.1g of erichrome black-T in 20 ml of ethyl alcohol (C_2H_5OH).

ii. Ammonia buffer, dissolve 6.75 g of ammonium chloride (NH_4Cl) in 58 ml of liquid ammonia and dilute to 100 ml with distilled water.

iii. Standard EDTA solution, dissolve 4 g of sodium salt of EDTA and 0.1 g of $MgCl_2$ in 800 ml of distilled water.

Procedure

Take 100 ml of the sample water in a conical flask. Add 1 ml of ammonia buffer and 3 drops of erichrome black T indicator. Titrate with standard EDTA solution till the colour changes from wine red to blue. Record the ml of EDTA solution used.

Calculation

$$\text{Total hardness} = \frac{\text{No. of ml of EDTA solution used x 1000}}{\text{ml. of sample}} \quad \text{mg/l}$$

Precautions

* Avoid air bubbling while immersing and filling DO bottle for the collection of water sample.

* Close the mouth of DO bottle with stopper inside water, *i.e.*, before taking it out of the water.

* Either freshly prepare the starch indictor or add into it one of the preservatives.

ESTIMATION OF NITROGEN

Nitrogen is a nutrient in the aquatic environment and it is necessary to sustain growth of aquatic organisms. It exists in several forms such as ammonia, nitrite and nitrate nitrogen.

Ammonia-nitrogen (NH^+_4-N)

It is the estimation of ammonium nitrogen (ionized, NH^+_4-N) by phenol-hypochlorite method. Ammonia reacts with phenol and hypochlorite in alkaline solution to produce indophenols blue and sodium nitroprusside is used to intensify the blue colour at room temperature. The dominance of NH_3-N (unionized) in water starts at the pH level above 9.75.

Reagents

1. **Phenol-nitroprusside buffer reagent:** Dissolve 30 g of sodium phosphate ($Na_3PO_4.12H_2O$), 30 g sodium citrate ($Na_3C_6H_5O_7.2H_2O$) and 3 g disodium salt of ethylene diamine tetra acetic acid (Na_2EDTA) in distilled water and make it up to one litre. Dissolve 60 g phenol (C_6H_5OH) and 0.2 g (200 mg) sodium nitroprusside ($Na_2(Fe(CN)_5NO).2H_2O$) in this solution, store in dark bottle and place in refrigerator.

2. **Alkaline hypochlorite solution:** Take 30 ml sodium hypochlorite (NaOCl having 10-14% available chlorine) to 400 ml 1 N NaOH and dilute with distilled water to one litre. Store in dark bottle and place in refrigerator. The 1 N NaOH can be prepared by dissolving 4 g NaOH in 100 ml distilled water.

3. **Stock solution for ammonia standard graph**

    a. **Dissolve** 0.943 g of ammonium sulphate $(NH_4)_2SO_4$ in one litre of distilled water. One ml of this solution contains 0.2 mg of NH^+_4 - N.

    b. **Dissolve** 3.821 g of ammonium chloride (NH_4Cl) in one litre of distilled water. One ml of this solution contains 1.0 mg of NH^+_4 - N.

Procedure

Take 25 ml of sample in Nesseler's tube and add 10 ml of phenol-nitroprusside buffer reagent, mix and promptly add 15 ml alkaline hypochlorite solution, cover Nesseler's tube and let the mixture stand in dark for one hour at room temperature. The colour gets changed to blue if ammonia is present. Measure the absorbance of standards and samples against a reagent blank at 635 nm.

Preparation of standard graph

Prepare the standard solution and make series of test tubes as per the following table and keep one blank.

Table 4: Quantification of ammonia–nitrogen (NH^+_4 - N)

SI. No.	Std. NH^+_4 - N solution (ml)	Distilled water (ml)	NH^+_4 - N (mg/L)	Optical density
1.	0	25.0	—	Blank
2.	0.5	24.5	0.10	
3.	1.0	24.0	0.20	
4.	1.5	23.5	0.30	
5.	2.0	23.0	0.40	
6.	2.5	22.5	0.50	
7.	3.0	22.0	0.60	
8.	3.5	21.5	0.70	
9.	4.0	21.0	0.80	

Nitrite – nitrogen (NO_2- N)

It is the estimation of nitrite nitrogen by formation of a reddish purple dye produced at pH 2.0 to 2.5 by coupling diazotized sulfanilic acid with naphthyl ethylenediamine dihydrochloride.

Reagents

1. **Sulfanilamide solution:** Dissolve 5 g sulfanilamide in 50 ml concentrated hydrochloric acid (HCl) and pour it slowly in distilled water and make the total volume 500 ml.

2. **Naphthyl ethylenediamine dihydrochloride:** Dissolve 500 mg naphthyl ethylenediamine dihydrochloride in 500 ml distilled water.

3. **Standard nitrite stock solution:** Take 0.246 g of anhydrous sodium nitrite ($NaNO_2$) dried at 110^0C and dilute it in one litre distilled water. One ml of this solution contains 50 ug of NO_2 - N.

4. **Working solution (for standard graph):** Take 10 ml of this stock solution and make it up to one litre with distilled water. One ml of this solution contains 0.5 ug of NO_2 - N.

Procedure

First take working solution as follows and prepare the standard graph. Now take 25 ml of sample as unknown and add 0.5 ml of sulfanilamide solution, wait 5–10 minutes, and add 0.5 ml of naphthyl ethylenediamine dihydrochloride solution. If the colour becomes pink take reading at 543 nm. Measure the NO_2-N from the standard graph against the optical density obtained.

Table 5: Quantification of nitrite – nitrogen (NO_2 - N)

Sl. No.	Std. NO_2 - N solution (ml)	Distilled water(ml)	NO_2-N (ug/L)	Optical density
1.	0	25	—	Blank
2.	1	24	0.5	
3.	2	23	1.0	
4.	3	22	1.5	
5.	4	21	2.0	
6.	5	20	2.5	

Nitrate - nitrogen (NO_3 - N)

It is the estimation of nitrate nitrogen by means of buffer reagent (phenol and sodium hydroxide solution) and reducing reagent (copper sulphate and hydrazine sulphate solution). The observation is made at 543 nm in spectrophotometer by using sulfanilamide and naphthyl ethylenediamine dihydrochloride producing a red azo-dye.

Reagents

1. **Buffer reagent**

   a. Phenol solution

   Dissolve 4.6 g of phenol in one litre of distilled water.

   b. Sodium hydroxide

   Dissolve 14.5 g of sodium hydroxide in one litre of distilled water.

2. **Reducing reagent**

   a. Copper sulphate

   Dissolve 100 mg of copper sulphate in one litre of distilled water.

   b. Hydrazine sulphate

   Dissolve 1.45 g of hydrazine sulphate in one litre of distilled water.

3. **Sulfanilamide solution**

   Dissolve 2.5 g sulfanilamide in 25 ml concentrate hydrochloric acid (HCl) and dilute it to 250 ml with distilled water. Place it in cool and dark place.

4. **Naphthyl ethylenediamine dihydrochloride**

   Dissolve 250 mg naphthyl ethylenediamine dihydrochloride in 250 ml with distilled water and store it in dark bottle at cool place.

5. **Acetone $(CH_3)_2CO$**

6. **Working solution**

   Take 2.589 g of potassium nitrate (KNO_3) in one litre distilled water. One ml of this solution contains 1.0 mg of NO_3-N (stock solution-1). Take one ml of this solution and make it to one litre with distilled water, one ml of this solution contains 1.0 ug of NO_3-N (stock solution-2).

Procedure

Pipette out 25 ml sample water, add 1 ml of buffer reagent (0.5 ml each, a and b) and shake well. Now add 1 ml of reducing agent (0.5 ml each, a and b), shake well and keep it in dark place for 24 hours at constant room temperature. After 24 hours, add 1 ml of acetone and then add 0.5 ml of sulfanilamide and after 2 minutes add 0.5 ml of diamine solution. If the colour becomes pink, take optical density after 15 minutes at 543 nm. Prepare the standard graph taking stock solution, get optical density and find out nitrate- nitrogen (NO_3-N) concentration.

Table 6: Quantification of nitrate – nitrogen (NO_3 - N)

Sl. No.	Std. NO_3 - N solution (ml)	Distilled water(ml)	NO_3 - N (ug/L)	Optical density
1.	0	25	———	Blank
2.	1	24	1.0	
3.	2	23	2.0	
4.	3	22	3.0	
5.	4	21	4.0	
6.	5	20	5.0	

ESTIMATION OF PHOSPHORUS (PO_4 - P)

Reagents

1. **Acid ammonium molybdate, $(NH_4)_6Mo_7O_{24}.4H_2O$)**

   Dissolve 10 g of ammonium molybdate in 70 ml of distilled water and then add 124 ml concentrate H_2SO_4 and now add 300 ml distilled water. Store in amber colour bottle.

2. **Stannous chloride, $(SnCl_2.2H_2O)$ fresh solution**

   Dissolve 0.5 g of stannous chloride in 2 ml of concentrate HCl acid and dilute to 20 ml with distilled water.

3. **Standard phosphate solution**

   Dissolve 4.387 g of oven dried potassium dihydrogen phosphate (KH_2PO_4) in one litre of distilled water (stock solution-1). One ml of this solution contains 1.0 mg of PO_4-P. Take one ml of this solution and make it to one litre with distilled water, one ml of this solution contains 1.0 ug of PO_4 - P (stock solution-2).

Procedure

Take 50 ml of sample and add 1 ml of ammonium molybdate solution, wait for 2-3 minutes, then add 0.2 ml of freshly prepared stannous chloride solution. Appearance of blue colour indicates the presence of phosphorus. Prepare the standard graph taking standard phosphate solution, get optical density and find out phosphorus (PO_4 - P) concentration at 690 nm.

Table 7: Quantification of phosphorus (PO_4 - P)

Sl. No.	Std. PO_4 - P solution (ml)	Distilled water(ml)	PO_4 - P (ug/L)	Optical density
1.	0	50	————	Blank
2.	1	49	1.0	
3.	2	48	2.0	
4.	3	47	3.0	
5.	4	46	4.0	
6.	5	45	5.0	

ESTIMATION OF SOIL TEXTURE

The sand, slit and clay percentage can be known either by soil texture test sieves or by soil particles gravitational settlement procedure in a measuring cylinder. The settlement of particles will express their gradation that can be observed and calculated as sand, silt and clay percentage.

ESTIMATION OF SOIL/SEDIMENT pH

Put 10 g dried sample in a 50 ml beaker. Add 20 ml distilled water and a pinch of barium sulphate ($BaSO_4$). Stir intermittently for 30 minutes and allow to settle. Pour the clear supernatant in a test tube and add two drops of universal indicator. Match the solution colour with corresponding pH colour (for dry sample). The pH for wet sample can be tested through soil pH cone device by pushing its lower part in the sample and in turn getting display of pH value at the upper dial portion.

ESTIMATION OF ORGANIC CARBON IN SOIL/SEDIMENT (Walkley and Black, 1934)

The sample is digested with a mixture of potassium dichromate and concentrate sulphuric acid, the organic carbon gets oxidized utilising the heat of dilution of sulphuric acid. The excess potassium dichromate is titrated with ferrous ammonium sulphate as redox titration.

$$2H_2Cr_2O_7 + 3C + 6H_2SO_4 = 2Cr_2(SO_4)_3 + 3CO_2 + 8H_2O$$

Reagents

1. **Standard 1 N $K_2Cr_2O_7$ solution**

    Dissolve 49.04 g $K_2Cr_2O_7$ in water and dilute to one litre.

2. **Standard 0.5 N ferrous ammonium sulphate, $Fe(NH_4)_2(SO_4)_2.6H_2O$ solution**

   Dissolve 196.1 g $Fe(NH_4)_2(SO_4)_2.6H_2O$ in 800 ml water containing 20 ml concentrate H_2SO_4 and dilute to one litre.

3. **Diphenylamine indicator, $(C_6H_5)_2NH$**

   Dissolve 0.5 g reagent grade diphenylamine in 20 ml water and 100 ml concentrate H_2SO_4.

4. **Phosphoric acid (H_3PO_4 – 85 %)**

5. **Sodium fluoride (NaF – 2 %)**

6. **Concentrate H_2SO_4**

7. **Concentrate H_2SO_4 plus silver sulphate (Ag_2SO_4) @ 1.25%**

Procedure

Take dried sample, grind it and pass through a 0.5 mm sieve. Place 1 g of the prepared soil sample in 500 ml conical flask. Add 10 ml of 1 N $K_2Cr_2O_7$ solution and 20 ml concentrate H_2SO_4. Mix thoroughly and allow the reaction to get it completed for 30 minutes. Dilute reaction mixture with 200 ml water and 10 ml H_3PO_4. Add 10 ml of NaF solution and 2 ml of diphenylamine indicator. Titrate the solution with standard 0.5 N ferrous ammonium sulphate solution to a brilliant green colour. A blank without soil is run simultaneously. If the titre value is lesser than 6 ml, repeat the determination taking 0.2 to 0.5 g of soil sample.

Calculation

$$\text{Organic carbon (\%)} = \frac{10(S-T)\times0.003}{S}\times\frac{100}{\text{Weight of soil (g)}}$$

Where

S = Ferrous ammonium sulphate solution required (ml) for blank.

T = Ferrous ammonium sulphate solution required (ml) for soil sample.

0.003 = Conversion factor, one ml of N $K_2Cr_2O_7$ is equivalent to 3 mg of carbon.

Note

1. High chloride content, as in the case of saline soils, interferes in the estimation process. It can be prevented by adding Ag_2SO_4 @ 1.25% to concentrate H_2SO_4.

2. Use of NaF along with H_3PO_4 gives a sharp end point.

3. In Analytical Chemistry, 'dissolve in water', denotes dissolve in distilled water.

CHAPTER 4

MONITORING OF AQUATIC BIOTA AND EUTROPHICATION

Aquatic life habitat is an essential part of aquatic community assemblages and life histories. The condition and type of habitat can envisage species diversity, growth rate and abundance. Monitoring aquatic habitat is crucial to assess the potential effects of any developmental activity on aquatic life. The monitoring can be used to assess differences between current habitat available and habitat available before, during and after developmental activities. Monitoring of aquatic biota and their habitat is needed to assess impacts on water quality and quantity discharged from any industrial activity or developmental project in short or long term. The monitoring of such biota is used to assess overall and site-specific biotic conditions.

Objectives

1. To determine the current type of aquatic life habitat available.
2. To assess the positive and negative changes in aquatic habitat over time.
3. To determine if changes in aquatic habitat are caused by developmental activity.
4. To develop mitigation measures to minimize potential negative effects to aquatic life habitat.

Plankton community

Plankton is a community including both plants and animals that consist of all those organisms whose powers of locomotion are insufficient to prevent them from being passively transported by currents.

Table 8: Size categories of plankton

The plankton have been categorized into following types on the basis of their size.

Sl. No.	Plankton	Size	Group
1.	Ultrananoplankton	< 2 μm	Free bacteria
2.	Nanoplankton	2 – 20 μm	Fungi, Flagellates, Diatoms
3.	Microplankton	20 – 200 μm	Most phytoplankton *spp.*, Foramini-ferans,Ciliates, Rotifers, Copepod nauplii
4.	Mesoplankton	200 μm – 2 mm	Cladocerans, Copepods, Larviceans
5.	Macroplankton	2 – 20 mm	Pteropods, Copepods, Chaetognathus
6.	Micronekton	20 – 200 mm	Cephalopods, Euphausiids
7.	Megaloplankton	> 200 mm	Scyphozoans, Thaliaceans

Group of plankton

Holoplankton group (true plankton) belongs to protozoa, cnidaria, ctenophore, nemeritinea, molluscs, annelida, crustacea, chaetonatha and chordata. However, from fisheries point of view, the plankton belonging to protozoa, crustacea and rotifer are much more important. Meroplankton refer to the organisms which only spend a portion of life cycle (usually the larval stage) within plankotnic form like sea stars and sea urchins.

PLANKTON SAMPLING DEVICES

Plankton samples can be collected by following three devices:

1. Sampling bottles

A wide mouth glass stoppered bottle of desirable volume is fitted on a graduated rope and sent down to the required depth, where the stopper is jerked open with messenger and the water is allowed to come in to the bottle. Kemmerer water bottle is such device. It is available in 3 capacities of 1000, 2000 and 3000 ml. Through these samplers, a known volume of water can be collected from a desired depth.

2. Plankton net

Plankton nets are comparatively much more suitable for qualitative estimation. There are many types of plankton nets available for different types of operations.

Normally for small sampling purposes, these are designed cone-shaped having a circular metal ring at one end and the plankton collecting tube on the other end. Generally bolting silk cloth with 40µ mesh size is used for zooplankton and 25µ mesh size for phytoplankton filtration.

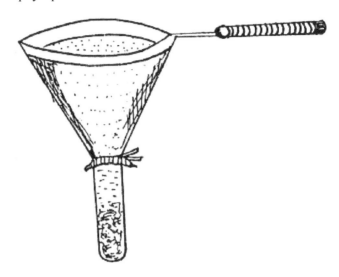

Fig. 2: Plankton collection net

3.   Pump and hose

Samples of plankton may also be collected by a suitable pump from any depth of water. Flexible inlet and outlet hose pipes are attached to the pump installed over a boat. The inlet hose is marked off in meters in order to read off the depth. A funnel fitted to the one end of the inlet hose prevents the escape of motile zooplankton. The diameter of the inlet hose pipe is generally in between 5 -10 cm depending on the suction power of the pump.

Aquatic insects

They form lesser than 4% of the total number of existing insect fauna. Nursery ponds designed to rear spawn of major carp are sometimes heavily populated during and after the rainy season with aquatic insects. They not only prey directly upon carp spawn and fry but also compete with the latter for food and available dissolved oxygen and hence their removal from these ponds is essential. Their best control measure is the netting of pond with mosquito net.

Benthos in aquatic ecosystem

The term benthos is widely referred to flora and fauna which are intimately associated with sediments in an aquatic system. Benthic environment represents bacteria, plants and animals including bottom living fishes from all phyla and their sizes widely varied. Benthic organisms are, in general, sessile and slow moving in nature. Seventy five percent of benthic animals live on firm substrates (rocks, corals *etc.*), twenty percent occur on sandy/muddy bottoms and only five percent of the total are planktonic. Development of morphological adaptations such as calcareous shells, elongated stalked and branched body, appendages with cilia and bristles and strong body musculature are characteristics of benthos. Many benthic animals within the sediment perform periodic vertical migration.

The size of benthos varied from microscopic to several orders of magnitude. Benthos are described on the basis of their position in the sediment with reference to surface and their size. The animals live within the sediment are called as "in-fauna". They move within the interstitial spaces or they build burrows or tubes. "Epi-fauna", which are either attached or move at or on the surface sediment. Faunal groups like amphipods, shrimps and scallops make extensive movements by swimming above the bottom. Based on size, they are classified into three categories, *viz.*, micro-fauna which vary from 1μ to 50μ (bacteria, protozoans, protophytes *etc.*), meiofauna which vary from 50μ - 1000 μ or 1mm (foraminiferans, copepods, nematodes, turbellarians, several larval forms *etc.*) and macro-fauna which are more than 1mm in size (all macroinvertebrates, macrophytes and demersal fishes).

Benthic organisms are reported from different kinds of substrates like hard, soft and coarse (sandy) bottom and also from organic matter and other macro debris. In general, they prefer firm substrate as compared to sandy or muddy bottom. Benthos are mostly either filter feeders or browsers or deposit feeders. Their feeding depends on particle size, quality abundance and choice preference. Their distribution is inversely proportional to depth on both horizontal and vertical realms. Their species diversity and composition vary widely with geographical area, climate, spatial and temporal differences, physico-chemical factors and other biological interactions. Species diversity, community structure, colonization, succession and the abundance in space and time of benthic communities are mainly decided by the following environmental factors, such as region, climate, temperature, salinity, water currents and circulation, depth, substratum, sediment grain size, oxidation – reduction state, dissolved oxygen, organic content and light.

Biological factors like

1) Food availability and feeding activities.

2) Prey-predator relationship and species removal.

3) Reproductive effects on breeding, spawning, dispersal and settlement.

4) Behavioural effects which induce movement and aggregation.

5) Growth and mortality.

Benthic flora

The benthic flora widely represent from phyto-benthos (algae and fungi) to macrophytes (aquatic weeds and grasses). In general, the flora are highly diversified and heterogeneous in size with dominance in selected habitats like pools, lakes, rivers, estuaries, intertidal salt marshes and shallow lagoons *etc*. Major components of phytobenthos are (a) cyanophysceans, (b) cryptomonads, (c) euglenoids, (d) dinoflagellates, (e) diatoms, (f) fungi and (g) photosynthetic bacteria. While major components of macrophytes are aquatic weeds and grasses, mangroves and salt marsh vegetations.

Macro and micro-algae and photosynthetic bacteria develop extensive populations in and on sediments and solid substrates depending on the availability of light intensity and oxidation – reduction condition. High biomass and production of algae are observed in salt marshes, intertidal sediments (80-200 gCm^{-2}y^{-1}) and in the sublittoral areas up to 20 to 30 m (< 80 gCm^{-2}y^{-1}) depth depending on the light penetrations. In sediment, photosynthetic activity is restricted to the upper 6-8 mm from the surface depending on the type of sediment and the wave length of light. Chlorophyll-a of the order of 100 mg/m^2 is common in intertidal sediments. In most of the cases about 2 to 4 mm from the surface sediment accounts for the compensation depth.

TYPES OF BENTHOS

Microbenthos

Micro fauna is mainly consisted of bacteria, ciliates and micro-flagellates. Bacteria are the most numerous and ubiquitous organisms present in sediments and their metabolic activities largely control the chemical nature of sedimentary environment. Bacteria in sediments are generally similar to those in water in antibiotic sensitivity, degradative and fermentative capability, and cation requirements. Common bacteria which occur in water- *Flavobacterium* and *Achromobacter* are often present in sediments. Both gram + ve and gram – ve cell types can be present and cells larger than bacteria can be isolated from water. Aerobic heterotrophs which oxidize organic compounds usually occur in surface sediments to depth determined by the penetration of dissolved oxygen. In anoxic conditions at deeper layer, anaerobic bacteria use various organic and reduced inorganic compounds as hydrogen acceptors. Bacteria capable of anaerobic respiration (fermentation) and chemosynthesis characterize these sediment layers. The bacterial biomass in the top 10 cm sediment varies from 5.5 gm^{-2} – 26.5 gm^{-2} depending on the sediment particle size. Bacteria account for more than 50% of the metabolism of benthic communities. In general, microorganisms are responsible for over 95% of community respiration.

Meiobenthos

Meiofauna include a wide assortment of metazoans with small elongated bodies adapted to an interstitial mode of life. They include ciliates, tardigrades, turbellarians, gastrotrichs, oligochaetes, archiannelids, harpacticoids, ostracods, nematodes, foraminiferans, rotifers and gnathostomulids. The meiofaunal community shows complex ecological properties, with predictable patterns of species composition and abundance in time and space and with recognizable patterns of resource partitioning. Micro flora and fauna form major source of food for the meiofaunal communities, although they use selectively the dissolved organic compounds to synthesize carbon. As compared to macrobenthos, meiobenthos account for about 10% of the total biomass and contribute equal production to that of macrofauna or greater than macrofaunal production. Meiofaunal representation is estimated at 1.6-1.7% of the total benthic production. About 16-20% of the meiofaunal production (average 20 $gCm^{-2}y^{-1}$) is utilized by macrofauna, while the rest is available for the mobile carnivores like fishes, birds and crabs, as well as used for within meiofaunal communities.

The larger metazoans over 500µ in size represent all the phyla form this group. Macrofauna mainly constitute three modes of feeding such as filter feeders (bivalves, sponges, ascidians, worms, barnacles *etc.*), browsers (amphipods, isopods, gastropods *etc.*) and deposit feeders (annelids, bivalves, gastropods, holothurians, crustaceans *etc.*). The preference of macrofauna to hard bottom and silty sand rather than muddy bottom is well recognized. Enumeration of population and community structure, biomass and species diversity of macrobenthos is much easier as compared to meiofauna and microfauna. Macrofaunal representation is estimated at 1.9-2.6% of the total benthic production and it accounts for very little to the total community representation of benthos. Macrofauna serve as a

Fig. 3: Common phytoplankton and zooplankton

major source of food for the demersal fishes. It is possible that the macrobenthos get about half of the sediment carbon and that many species are able to transform it to macrobenthos production with efficiency around 40%. In that case, an overall benthic transformation efficiency of 20% may be realizable and yield to demersal fish of 5 $gCm^{-2}y^{-1}$ be well within the capacity of most benthic communities.

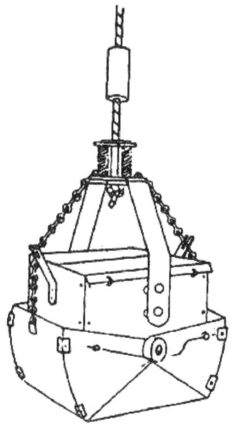

Fig. 4: Ekman's grab sediment sampler

ROLE OF BENTHOS

Benthos play a vital role in the aquatic food chain, bioturbation and in the recycling of essential life sustaining elements like, C, N and P in an aquatic ecosystem. The settled organic matter from the water column is effectively consumed and converted into invertebrate benthic biomass; dissolved organic matter and inorganic nutrients by benthic organisms. The nutrients released from the sediments due to bacterial degradation of organic matter, diffuse and disperse rapidly into the

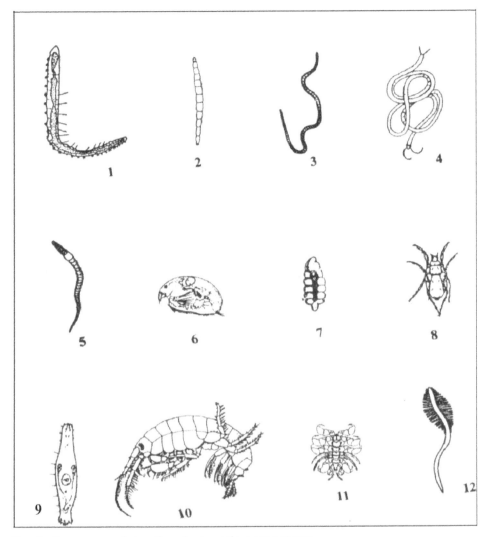

Fig. 5: Common meiobenthos in aquatic ecosystems
1. Nais, 2. Culicoides, 3. Lumbriculus, 4. Polygordius, 5. Lumbricilius, 6. Alona, 7. Cryptonisus, 8. Halacarellus, 9. Macrostomida, 10. Gammarus, 11. Cyamus, 12. Branchiura.

overlying water and influence the primary production which in turn triggers the zooplankton production in an aquatic environment. There is certain amount of evidence that significant quantum of nitrogen and phosphorus is released from sediments in organic form which is directly utilized by selected microfauna and meiofauna for the organic synthesis. Bacteria play a vital role in the process of decomposition of organic matter, nutrient recycling, in the energy cycle of benthic ecosystem and as a source of food for benthic animals. Bacteria are responsible

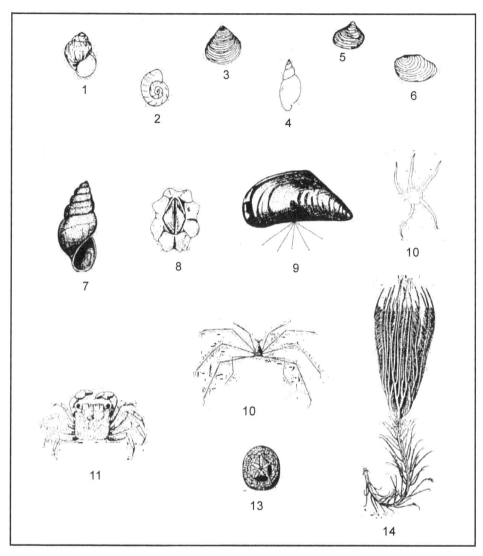

Fig. 6: Common macrobenthos in aquatic ecosystems
I. Viviparus, 2. *Gyraulus*, 3. *Corbicula*, 4. *Lynnaea*, 5. *Pisidium*, 6. *Lamellidens*,
7. *Potamopyrgus*, 8. *Eliminus*, 9. *Mytilus*, 10. *Ophioglyphi*, 11. *Sesarma*, 12. *Latreillie*,
13. *Clypeaster*, 14. *Metacrinus*

for more than 50% of the metabolism of benthic communities. In brief, they are responsible for creation and protection of aquatic food chain. They are the major source of food for meiobenthos and macrobenthos which in turn serve as an important source of food for demersal fishes. Since, the demersal fishery contributes about 30 to 50% of the total fishery potential of an area; the benthic production plays a major role in deciding the demersal fishery potential.

The macrobenthos create bio-turbidation during their movements and feeding activities which condition the sediments for meiofauna and microfauna and as a stimulant of nutrient regeneration. Many deposit feeders ingest anaerobic sediments and transfer them to surface layer where they get oxidized. This also helps transfer of bacteria and organic matter from deeper reduced layer to surface oxidized zone. Simultaneously, this also aids the transport of well oxygenated water from the surface to deeper zones. Benthos due to their differential tolerance have been considered as the best indicator organisms of environmental stress or aquatic pollution. Benthic macroinvertebrates are widely used in 'biomonitoring' programmes both as surveillance and compliance in order to assess the health of the aquatic environment.

In the aquaculture practices, benthos play a significant role in the assessment and monitoring of culture ponds. If the benthic ecosystem of the culture pond is properly maintained, the benthos form an additional source of natural food for the culture species like shrimps which invariably feed on benthic organisms. The benthic ecosystems of coral reef, mangroves, intertidal zones and mudflats serve as a good feeding, breeding, spawning and nursery grounds for many marine organisms of economical importance, variety of migratory and resident birds, fishes, reptiles and sea mammals.

EUTROPHICATION

Eutrophication of water bodies is an important environmental issue related to aquaculture. Its impact on aquaculture is both direct and indirect. It affects aquatic ecosystem, when it occurs in culture pond even may cause mass mortality of aquatic organisms, due to gill clogging and depletion of dissolved oxygen during dark hours. In other ways, nutrients released from intensive aquaculture ponds, generate and accelerate eutrophic condition in natural water bodies, which is an environmental safety issue. It may be caused by excessive fertilization, feeding and other nutrients inputs like agriculture runoff, sewage pollution and other allochthonous sources *etc*. Eutrophication is an increase in chemical nutrients, the typical compounds containing nitrogen and phosphorus in an ecosystem. Eutrophication generally promotes excessive chlorophyll bearing biomass growth and favours certain weedy species over others and is likely to cause severe deterioration in water quality. Precisely, "Eutrophication is an accelerated growth of chlorophyll bearing biomass caused by the over enrichment of water by nutrients, especially compounds of nitrogen and/or phosphorus and inducing an undesirable disturbance to the balance of organisms present in the water and its quality". It has been observed that higher concentration, more than 250 ppm, of carbonates and bicarbonates accelerates weed infestation in aquatic bodies rather than algal bloom. It means higher concentrations of nitrogen and phosphorus cause algal eutrophication while weed infestation is the effect of the large amount of carbonate and bicarbonate.

Table 9: Characteristics of eutrophication

Sl. No.	Trophic level	Total P (µg/L)	Chlorophyll-a (µg/L)	Secchi depth (m)
1.	Ultraoligotrophic	< 5	< 1	> 12
2.	Oligomesotrophic	50-100	1- 2.5	> 6
3.	Mesoeutrophic	100-500	2.5-8	6-3
4.	Eutrophic	500-1000	8-25	3-1
5.	Hypereutrophic	> 1000	> 25	< 1

Negative role of unwanted weeds

Presence of weeds in aquaculture ponds restricts the movement of fish and their ability to chase and hunt for food. Therefore, weed infestation in aquaculture ponds is not desirable. Weeds provide shelter to unwanted fishes which hunt upon and compete for food with cultivated species. Presence of weeds not only reduces water holding capacity of the system but also accelerates the rate of water evaporation and process of siltation. Water circulation, aeration and light penetration are the other parameters which adversely get affected by weed infestation. Thus, all these factors are the causes for the reduction of fish yield. Comparatively, the root system in grasses is rather well developed than weeds.

Control of eutrophication

Aluminium sulphate ($Al_2(SO_4)_3.9H_2O$) or alum yields sulphate and aluminium ions, and aluminium hydroxide complexes such as $Al(OH)_2^+$ and $Al(OH)_3^+$; when they get dissolved in water, act as the excellent coagulant and hence reduction in the levels of eutrophication and turbidity occurs. In unfertilized ponds, alum treatment will lead to reduce phosphorus and plankton production. Researchers found that alum-induced reductions in hypolimnetic phosphorus decreased the frequency and severity of phytoplankton blooms, with treatment effectiveness lasting up to 14 years. However, effects on phytoplankton were offset by increase in macrophyte growth and distribution after treatment. In the absence of neutralizing agents, alum concentration can increase, poisoning aquatic organisms and lowering pH to unacceptable values in inadequately buffered systems. Inactivation of sediment phosphorus is usually employed following controls on excessive external loads. For best effect, alum should be distributed evenly over the pond surface by dissolving in water and spraying. Alum is more effective than ferric sulphate, hydrated lime or gypsum in removing clay turbidity in laboratory experiments. As much as 84-97% of turbidity may be removed in 24 to 48 hours if waters have total alkalinity in the range of 275 to 350 mg/liter; liming may be required after alum application to prevent depression of alkalinity and pH which could be caused by the application of alum. The quantum of dose can be ascertained by jar test procedure.

Lime (CaO, $CaCO_3$, $Ca(OH)_2$) application has been used primarily as a lake rehabilitation technique for limiting algal growth by controlling phosphorus availability in the water column and its release from the sediment. Lime application supersaturates the water column with Ca^{2+}, which can result in P^{3-} precipitating out of solution as hydroxyapatite. As a deposited layer on the sediment, it can absorb additional phosphorus (at pH > 8), preventing it from diffusing into the water column for algal uptake. Lime can also induce phytoplankton flocculation, which causes cells to sink below the euphotic zone, thus further reducing phytoplankton biomass. Copper sulphate ($CuSO_4.5H_2O$) can also be used to control algal bloom eutrophication @ alkalinity x 0.01ppm.

Bioindicators

Trace metals, organochlorines and other pollutants may be accumulated by aquatic biota in their body to levels far above those found in the surrounding environment. This phenomenon leads to the possibility of using selected organisms to monitor the levels of pollutants in water bodies. These selected organisms to monitor the pollutants levels are known as bioindicators.

CHAPTER 5

DESIRABLE RANGES OF FRESH, BRACKISH AND SEAWATER FOR AQUACULTURE

A vailable raw water quality at source may differ from the aquaculture operations' water quality requirement; it can be known by the following table.

Table 10: Favourable ranges of water quality parameters for stress free fish rearing and culture in three types of water bodies:

Sl. No.	Parameter	Freshwater	Brackish water	Seawater
1.	Colour (colour unit)	Plankton turbidation, greenish hue and < 100 colour units	Plankton turbidation, greenish hue and < 100 colour units	Plankton turbidation, greenish hue and < 100 colour units
2.	Transparency (cm)	20-35	25-35	25-35
3.	Light/illuminance (lux)	5000-20,000	5000-20,000	5000-20,000
4.	Turbidity*			
	Clay turbidity (ppm)	< 35	< 35	< 35
	Clay turbidity (NTU)	< 100	< 100	< 100
5.	Solids (mgl^{-1})			
	a. Total dissolved solids (TDS)	< 500	> 500	> 500
	b. Total suspended solids (TSS)	30-200	25-200	25-200

[Table Contd.

Contd. Table]

Sl. No.	Parameter	Freshwater	Brackish water	Seawater
6.	Temperature (^{0}C)			
	a. Tropical climate	25-32	25-32	25-32
	b. Temperate climate	10-18	10-18	10-18
7.	pH**	7.5-9.0	8.0-8.5	8.2-8.5
8.	Hardness (mgl^{-1})	50-200	> 275	> 1700
9.	Alkalinity (mgl^{-1})	50-275	> 100	> 100
10.	Chlorides (Cl: mgl^{-1})	30-70	> 500	> 500
11.	Salinity (ppt)	< 0.5	10-30	> 30
12.	Dissolved oxygen (O_2: mgl^{-1})	5-10	5-10	5-10
13.	Total dissolved free carbon dioxide (CO_2: mgl^{-1})	< 3	< 3	< 3
14.	Ammonia nitrogen			
	a. Unionized (NH_3-N: mgl^{-1})	0-0.1	0-0.1	0-0.1
	b. Ionized (NH_4^+-N: mgl^{-1})	0-1.0	0-1.0	0-1.0
15.	Nitrite nitrogen (NO_2-N: mgl^{-1})	0-0.5	0-0.5	0-0.5
16.	Nitrate nitrogen (NO_3-N: mgl^{-1})	0.1-3	0.1-3	0.1-3
17.	Total nitrogen (mgl^{-1})	0.5-4.5	0.5-5.0	0.5-5.0
18.	Total phosphorous (mgl^{-1})	0.05-0.4	0.05-0.5	0.05-0.5
19.	Potassium (K: mgl^{-1})	0.5-10	> 0.5	> 0.5
20.	Calcium (Ca: mgl^{-1})	75-150	75-400	75-400
21.	Magnesium (Mg: mgl^{-1})	20-200	200-1350	> 1350
22.	Sulphate (SO_4: mgl^{-1})	20-200	200-885	> 885
23.	Silicates (SiO_2: mgl^{-1})	4-16	> 5	> 5
24.	Iron (Fe: mgl^{-1})	0.01-0.3	0.01-0.3	0.01-0.3
25.	Manganese (Mn: mgl^{-1})	0.001-0.002	0.002-0.02	0.002-0.02
26.	Zinc (Zn: mgl^{-1})	0.002-0.01	0.002-0.01	0.002-0.01
27.	Copper (Cu: mgl^{-1})	0.003-0.005	0.003-0.005	0.003-0.005
28.	Cobalt (Co: mgl^{-1})	< 0.003	< 0.003	< 0.003
29.	Biochemical oxygen demand (mgl^{-1})	< 10	< 15	< 15
30.	Biochemical oxygen demand (BOD:Kg/ha/day for sewage fishery)	150	100	75
31.	Chemical oxygen demand(COD: mgl^{-1})	< 50	< 70	< 70
32.	Hydrogen sulfide (H_2S: mgl^{-1})	< 0.002	< 0.003	< 0.003
33.	Fluoride (F: mgl^{-1})	< 1.5	< 1.5	< 1.5

[Table Contd.

Contd. Table]

Sl. No.	Parameter	Freshwater	Brackish water	Seawater
34.	Residual chlorine (mgl^{-1})	< 0.003	< 0.003	< 0.003
35.	Primary productivity (mg C/m^3/day)	1000-3000	1000-2500	1000-2500
36.	Plankton (ml/100 litre)	1-2	1	1
37.	Chlorophyll - a (µgl^{-1})	20-275	20-250	20-250
38.	Redox - potential (Eh: mv water)	350-400	350-400	350-400
39.	Redox - potential (Eh: mv sediment)	250-400	200-300	200-300
40.	Organic carbon in sediment (%)	0.50 – 2.75	0.50 – 2.75	0.50 – 2.75

* One NTU is about 2.5 to 3.0 ppm.

** More than 7 pH of water is necessary for molluscs' spp. culture as the calcium carbonate ($CaCO_3$) shells of molluscs can begin to get disintegrated and dissolved under acidic conditions.

CHAPTER 6

GUIDELINES

To streamline the aquaculture practices there are following guidelines and accepted standards for more efficient and profitable aquaculture.

Table 11: Maximum permissible residual levels for fish and fishery products (CAA, 1999)

Sl. No.	Substance	Permissible level (ppm)
A.	**Antibiotics and other pharmacologically active substances**	
1.	Chloramphenicol	Nil
2.	Nitrofurans including: Furaltadone, Furazolidone,Furylfuramide, Nifuratel, Nifuroxime, Nifurprazine,Nitrofurantoin, Nitrofurazone	Nil
3.	Neomycin	Nil
4.	Nalidixic acid	Nil
5.	Sulphamethoxazole	Nil
6.	*Aristolochia spp.* and preparations there of	Nil
7.	Chloroform	Nil
8.	Chlorpromazine	Nil
9.	Colchicine	Nil
10.	Dapsone	Nil
11.	Dimetridazole	Nil
12.	Metronidazole	Nil
13.	Ronidazole	Nil
14.	Ipronidazole	Nil
15.	Other nitroimidazoles	Nil
16.	Clenbuterol	Nil
17.	Diethylstilbestrol (DES)	Nil

[Table Contd.

Contd. Table]

Sl. No.	Substance	Permissible level (ppm)
18.	Sulfonamide drugs (except approved)	
	Sulfadimethoxine, Sulfabromomethazine and Sulfaethoxypyridazine)	Nil
19.	Fluroquinolones	Nil
20.	Glycopeptides	Nil
21.	Tetracycline	0.1
22.	Oxytetracycline	0.1
23.	Trimethoprim	0.05
24.	Oxolinic acid	0.3
B.	**Substances having anabolic effect and unauthorized substances**	
1.	Stilbenes, stilbene derivatives and their salts and esters	
2.	Steroids	
C.	**Veterinary drugs**	
1.	Antibacterial substances, including quinolones	
2.	Anti-helminthic	
D.	**Other substances and environmental contaminants**	
1.	Organochlorine compounds including PCBs	
2.	Mycotoxins	
3.	Dyes	
4.	Dioxins	
E.	**Pesticides**	
1.	BHC	
2.	Aldrin	
3.	Dieldrin	
4.	Endrin	
5.	DDT	
F.	**Heavy metals**	
1.	Mercury	
2.	Cadmium	
3.	Arsenic	
4.	Lead	
5.	Tin	
6.	Nickel	
7.	Chromium	

Table 12: Standards for treatment of wastewater discharged from the aquaculture farms, hatcheries, feed mills and processing units (CAA, 2002)

Sl. No.	Antibiotics and other pharmacologically active substances	Final discharge point	
		Coastal waters	Creek or estuarine courses when the same inland water courses are used as water source and disposal point
1.	pH	6.0 – 8.5	6.0 – 8.5
2.	Suspended solids (mg/l)	100	100
3.	Dissolved oxygen (mg/l)	Not lesser than 3	Not lesser than 3
4.	Free ammonia (as NH_3 –N, mg/l)	1.0	0.5
5.	Biochemical oxygen demand (5 days @ 20^0 c, mg/l)	50	20
6.	Chemical oxygen demand, mg/l	100	75
7.	Dissolved phosphate as P (mg/l)	0.4	0.2
8.	Total nitrogen as N (mg/l)	2.0	2.0

Table 13: List of antibiotics and other pharmacologically active substances banned for shrimp aquaculture (CAA, 1999)

Sl. No.	Antibiotics and other pharmacologically active substances
1.	Chloramphenicol
2.	Nitrofurans including: Furaltadone, Furazolidone, Furylfuramide, Nifuratel, Nifuroxime, Nifurprazine, Nitrofurantoin, Nitrofurazone
3.	Neomycin
4.	Nalidixic acid
5.	Sulphamethoxazole
6.	*Aristolochia spp.* and preparations thereof
7.	Chloroform
8.	Chlorpromazine
9.	Colchicine
10.	Dapsone
11.	Dimetridazole
12.	Metronidazole
13.	Ronidazole

[Table Contd.

Contd. Table]

SI. No.	Antibiotics and other pharmacologically active substances
14.	Ipronidazole
15.	Other nitroimidazoles
16.	Clenbuterol
17.	Diethylstilbestrol (DES)
18.	Sulfonamide drugs (except approved Sulfadimethoxine, Sulfabromomethazine and Sulfaethoxypyridazine)
19.	Fluroquinolones
20.	Glycopeptides

Do's and Don'ts for Culture of SPF *L. Vannamei* (As per CAA Guidelines)

1. The farm should have been registered with the Coastal Aquaculture Authority (CAA).

2. Permission of CAA to take up SPF *L. vannamei* culture should have been obtained.

3. Biosecurity requirements include:

   a. Farm to be fenced (including crab fencing),

   b. Provide reservoir for water intake,

   c. Provide bird scares/bird netting,

   d. Separate implements for each pond.

4. Should employ trained/experienced personnel in management of biosecurity measures.

5. Irrespective of size, farms should have an Effluent Treatment System (ETS).

6. Quality of wastewater should conform to the standards prescribed by CAA. Wastewater to be retained at least for two days and should be chlorinated and dechlorinated before its release into drainage system. Water exchange should be avoided, if possible.

7. Do not culture SPF *L. vannamei* if neighbouring farms culture native species of shrimp, crab *etc*.

8. Only tested and certified seed produced by CAA authorized hatcheries should be used. Keep record of the name and address of hatchery, number and date of registration of hatchery and date and quantity of seed procured. PL for stocking should be selected using standard morphological and health check.

9. Record the quantity of shrimp produced and sold and name and address of the processor to whom sold and report it to CAA in the prescribed proforma.

10. Use only pelleted feed manufactured by reputed companies meant for SPF *L. vannamei* in the prescribed quantity to minimize feed wastage. When necessary, reasonable reduction in feeding must be performed in order to improve the water quality. No banned chemicals/antibiotics should be used in the feed or in any other form.

11. Depth of the ponds should be maintained at a level not lesser than 1.5 m.

12. If the stocking density exceeds 5 pieces/m^2, aerators may be used to keep up the level of oxygen requirement. Dissolved oxygen content of pond water should be maintained above optimum level (above 5 mg/l) throughout the culture period.

13. Disinfecting protocol for personnel and implementations should be strictly adhered to.

14. Shrimp health and amount of bacteria (vibrios) in water column and water quality should be regularly monitored.

15. Water quality parameters such as pH, alkalinity, dissolved oxygen, total ammonia, unionized ammonia, nitrite, nitrate, total nitrogen, phosphorus, BOD and COD should be regularly monitored.

16. Outbreak of disease should be reported immediately and must be treated and contained within the pond in order to prevent spread of disease.

17. Area surrounding the farm should be kept clean from garbage and other farm wastes.

18. Storage for feed and farm equipments should be in good condition and kept away from any potential carrier animals.

19. Harvesting should be planned in advance and proper care should be taken for icing the product immediately to maintain product freshness and sanitation.

20. Record of farm management practices including feeding, water quality, chemical use and disease treatment, therapeutic agent *etc.* should be maintained as a routine practice to provide up-to-date information to the Coastal Aquaculture Authority, Chennai (India).

CAA: Coastal Aquaculture Authority

SPF: Specific pathogen free

Table 14: Water Quality Criteria by Central Pollution Control Board (Ministry of Environment, Forest and Climate Change, Govt. of India

Designated-Best-Use	Class of water	Criteria
Drinking water source without conventional treatment but after disinfection	A	• Total coliform organisms MPN/100 ml shall be 50 or less • pH between 6.5 and 8.5 • Dissolved oxygen 6 mg/l or more • Biochemical oxygen demand 5 days @20°C, 2 mg/l or less
Outdoor bathing (organized)	B	• Total coliform organisms MPN/100ml shall be 500 or less • pH between 6.5 and 8.5 • Dissolved oxygen 5 mg/l or more • Biochemical oxygen demand 5 days @20°C, 3 mg/l or less
Drinking water source after conventional treatment and disinfection	C	• Total coliform organisms MPN/100ml shall be nil • pH between 6 to 9, dissolved oxygen 4 mg/l or more • Biochemical oxygen demand 5 days @20°C, 3 mg/l or less
Propagation of wild life and fisheries	D	• pH between 6.5 to 8.5, dissolved oxygen 4 mg/l or more • Free ammonia (as NH_3-N), 1.0 mg/l or less
Irrigation, industrial cooling and controlled waste disposal	E	• pH between 6.0 to 8.5 (Electrical conductivity at 25°C, micro mhos/cm, Max. 2250) • Sodium absorption ratio, Max. 26 • Boron, Max., 2 mg/l

CHAPTER 7

ZERO WATER EXCHANGE

The zero water exchange concept

In a zero water exchange system, no water is discharged from the system and no additional water, except for evaporation, is added after the system is initially filled. The use of zero-exchange systems has become a viable alternative to traditional ponds methods of intensive aquaculture production. The concept has been applied to indoor tank based production systems, outdoor intensive systems and large volume marine aquariums. Today, zero water exchange, the ecological recirculating aquaculture system has forged a path towards industry – leading technology.

Zero-water exchange aquaculture

The fish catch from natural resources has been getting depleted every year mainly due to two main causes. First, the burgeoning aquatic pollution problem and second, the over exploitation of the aquatic resources. The exploitation is not only of adult stages but also of early stages of life cycle of adult ones. Therefore, the aquatic life and environment is threatened up to an extent that the number of dead – points/ zones (no availability of dissolved oxygen) are increasing in the aquatic systems the world over at an alarming rate. In the last two decades, the aquaculture developed with a very fast pace due to high demand in the international market coupled with increasing human population. But this development could not last longer as the amplitude of range of development was ecologically imbalance which resulted the disease out-break the world over. To get rid of these problems, the ecologically balanced zero water exchange system came in to existence and it is now fully scientifically and technically proven technology for its application in enhanced aqua-production systems.

Characteristics of zero water exchange

Ecological recirculatory aquaculture system keeps ecological balance away from chemical way,

Preventing sudden and unnecessary infections,

Economical energy cost,

Increase in stocking density and

Higher survival.

Components for open systems

1. Physical filter

2. Biological multi-layer aeration filter

3. Ecological building system

Table 15: Components for closed systems

Scientific	Technical	Engineering
The problems arising out of the causes like:	The technical know-how on the aspects like:	The engineering installations like:
Incidental/Accidental	Instrumentation	Sand filter
Routine	Mechanical	Biofilter
Shelf-life		Protein skimmers
Chemical		Pumps
Biological		Aeration machinery and air distribution system
		Ozonizer and scientific application of ozone into the system

Laboratory upgradation

The inputs needed for:-

● Electronic gadgets

● Titrimetric

Table 16: Favourable ranges of water quality parameters for open and closed zero water exchange systems

Sl. No.	Parameter	Unit	Favourable range			
			Open system		Closed system	
			Fresh water	Saline water	Fresh water	Saline water
1.	pH	–	7.5-9.5	8.2-8.6	7.0-9.0	8.0-8.5
2.	Dissolved oxygen	ppm	5-10	5-10	5-6	5-6
3.	Carbondioxide	ppm	< 3	< 3	< 4	< 4
4.	Salinity (species specific)	ppt	< 0.5	0.5-37.0	< 0.5	0.5-37.0
5.	Ammonia nitrogen,					
	Ionized ($NH_4^+ - N$)	ppm	< 1.0	< 1.0	< 1.2	< 1.2
	Unionized ($NH_3 - N$)	ppm	< 0.10	< 0.10	< 0.12	< 0.12
6.	Nitrite Nitrogen ($NO_2 - N$)	ppm	< 0.50	< 0.50	< 0.60	< 0.60
7.	Nitrate Nitrogen ($NO_3 - N$)	ppm	< 5.0	< 5.0	< 5.5	< 5.5
8.	Phosphorus ($PO_4 - P$)	ppm	0.05-0.50	0.05-0.50	0.06-0.60	0.06-0.60

Action required

The complete set-up of water treatment system needs the regular servicing along with constant monitoring of the water parameters as mentioned in the table 16.

The following parameters are to be filled in the laboratory log book –

● Temperature

● pH

● Dissolved oxygen

● Carbon dioxide

● Salinity/hardness/alkalinity

● Ammonia-nitrogen

● Nitrite-nitrogen

● Nitrate-nitrogen

● Phosphorus

It is necessary to check all the above mentioned parameters regularly to run the water quality management system smoothly and effectively and in turn it will entrust the pre-warning signals for the disease outbreak or the mortality of the fishes.

Some of the above parameters can be automated in the running water cycle system where the sensors can be fitted in the pipe line and the measured values can be recorded in the computer; there by constantly monitoring the water recycle system.

Considering the quantity and the growth rate of each species of fish, the bioload and the water change cycle have to be closely monitored and maintained appropriately.

Use of good quality vitamin enriched fish food and feeding cycle should be monitored depending on the food habits of the types, quality and total quantum of fish stock.

STANDARD OPERATION PROCEDURES

Monitoring

1. Check positions of the light.
2. Check level of protein skimmer.
3. Check level of water in the balancing tank.
4. Check all equipments and pumps, ensure all in working conditions.
5. Check aeration/ozone distribution lines.
6. Quarantine any diseased fish.

Laboratory work

1. Check water parameters regularly.
2. Backwashing of the sand filters.
3. Check aeration/ozone system regularly.

   Record book: Maintain log book.

Process components

A. **Pressure sand filter:** Water is first filtered by pressure sand filter unit for removal of suspended material such as sediments, organisms, colloids, particulate matter, fiber, dust particles and turbidity.

B. **Granular activated carbon filter:** Activated carbon filter unit is used to remove any additional chlorine and organic matter from water without release of carbon particles. It also removes colour and odour from the water; thus acts as decolourisation and deodourisation agent.

C. **Micron filtration unit:** These are pressure vessels fitted internally with polypropylene filter cartridge elements of 5 micron and 20 micron rating, which can remove micron size particles. The filter elements need to be replaced after every six months intervals.

D. **Osmosis pressure system: (booster pump):** A high pressure booster pump is provided for supplying the feed water to the RO membrane unit at high pressure of 12 to 14 kg/cm.

E. **RO membrane element unit:** A high rejection (98%) thin film composite RO membrane is used to reject wide spectrum of impurities including total dissolved solids (TDS), bacteria and viruses. The pure water is further treated in the UV sterilizing unit, while the rejected water (brine) is automatically flushed down to the drain. The RO membrane element under proper care and use needs to be replaced after regular intervals.

F. **UV sterilizing unit:** A UV sterilizing unit is provided to disinfect the treated water from water borne pathogens, bacteria, viruses *etc.*, which may have passed through the RO membrane unit. The water is thus completely safe before put to use.

G. **Final polishing filter:** This is the final polishing stage before the product water is stored in the storage tank or consumed further in process. It is used to remove dissolved gases, bad taste and odour from the product water. The filter element needs to be replaced after every one year interval depending on the quality of available raw water.

The problems and their solutions in zero water exchange system

The problems arise due to the decrease in water pH, alkalinity, dissolved oxygen, and the increase in ammonia nitrogen, nitrite nitrogen and carbondioxide. The fluctuations of these parameters beyond favourable ranges cause the deterioration of water quality in the system which can lead to bacterial, fungal and protozoan ciliate infection in turn the disease outbreak and ultimately fish/shrimp mortality. Under such deteriorated environment, to have a turn towards optimal water quality condition in the system, following ones are the suggestions.

1. The pH and alkalinity can be controlled by the application of sodium bicarbonate ($NaHCO_3$) @ 10 ppm (10 g/1000 litre).

2. The application of hydrogen peroxide (H_2O_2) @10 ppm (10 ml/1000 litre) can solve the problem of dissolved oxygen deficiency and control of ammonia nitrogen, nitrite nitrogen and carbondioxide.

3. The bacterial and fungal infection can come down with the application of copper sulphate ($CuSO_4.5H_2O$) @ 0.5 ppm (500 mg/1000 litre).

4. The protozoan ciliate infection can be removed by applying potassium permanganate ($KMnO_4$) @ 0.5 ppm (500 mg/1000 litre).

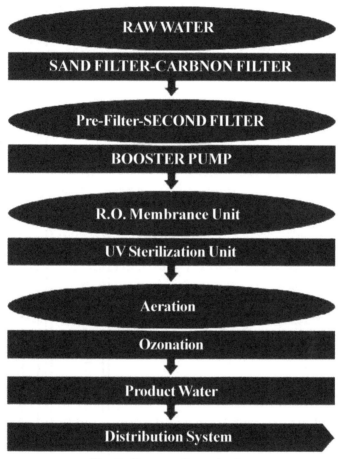

Fig. 7: Process flow charts

Conclusion

The biofloc keeps microbial diseases away and zero-water exchange makes it ecofriendly, while recirculatory aquaculture systems are used where water exchange is limited, but comparatively more capital intensive. A traditional filter system filters non-dissolved impurities and suspended solids out of the water by physical methods. However, fish and shellfish excrete toxic ammonia, which needs to be transformed to nitrite-nitrogen and nitrate-nitrogen through oxidation by nitrifying bacteria to prevent stress, health troubles and mortality. Traditionally,

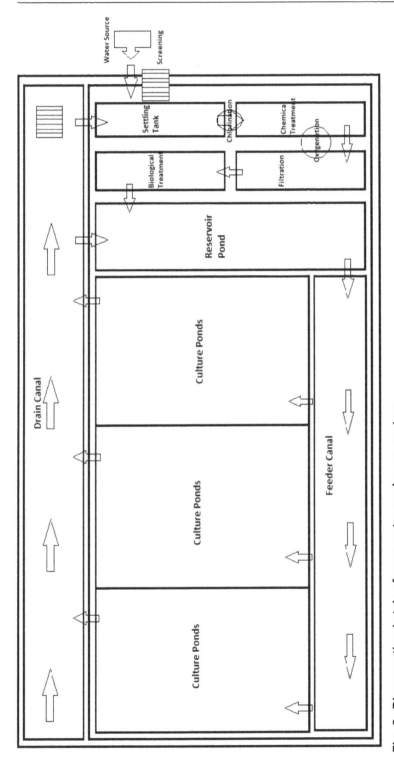

Fig. 8: Diagramatic sketch of zero water exchange system

in aquaculture farms, dilution of ammonia and nitrite-nitrogen is done by water exchange but this may affect the fish and shell fishes' immune system function and consequently the survival rate. The threat to fish and shellfishes be solved if ammonia nitrogen and nitrite-nitrogen can quickly be transformed into non-toxic nitrate-nitrogen, which is the best nitrogen fertilizer that can be utilized by bacteria, phytoplankton and zooplankton thus enhance stable water quality and allow for a zero water exchange. The need based zero water exchange systems can be designed and fabricated taking into consideration the above scientific and technical inputs.

CHAPTER 8

SCREENING, COAGULATION, SEDIMENTATION AND LIMING

Screening

Screens are generally provided in front of the pumps or the intake works so as to exclude the large size particles or floating material. Coarse screens generally called (trash racks) are sometimes placed in front of fine screens. Coarse screens consist of parallel iron rods placed vertically or at a slide slope at about 2-10 cm centre to centre. Medium screens have the spacing between the bars as 6 mm to 40 mm. The fine screens are made up of thin wire or perforated metal with openings of 1.5 mm to 4.0 mm size. The material which is collected on the up streams side of the screens is removed either manually or mechanically.

Coagulation

It refers the collection of minute solid particles dispersed in a liquid medium into a larger mass. Raw water with 30 ppm turbidity can be treated by filtration without any pre-treatment. If turbidity exceeds more than 30 ppm that too if it has become undesirable for specific cultivable organisms, the alum can be applied as per jar test procedure observing out the optimum dose per litre of sample water. Commercial alum containing combination of $Al_2 (SO_4)_3.18H_2O + Al_2 (SO_4)_3.9H_2O$ can be used as a coagulant. Alum reacts with natural alkalinity and produces a floc of aluminium hydroxide along with suspended or colloidal impurities due to agglomeration process. So, by the application of coagulant alum, suspended impurities can be settled out more quickly.

- $Al_2(SO_4)_3.\ 9H_2O + 3Ca(HCO_3)_2 = 2Al(OH)_3 + 3CaSO_4 + 9H_2O + 6CO_2$
- $Al_2(SO_4)_3.\ 9H_2O + 3MgCO_3 = 2Al(OH)_3 + 3MgSO_4 + 6H_2O + 3CO_2$

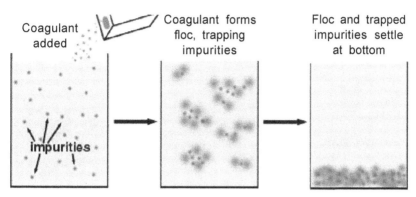

Fig. 9: Coagulation and flocculation

Sedimentation

Most of the suspended impurities present in the water do have a specific gravity greater than that of water (*i.e.*, 1.0). In still water, these impurities will, therefore, tend to settle down under gravity. Although in normal raw supplies, they are kept in suspension because of turbulence in water. Hence, as soon as the turbulence

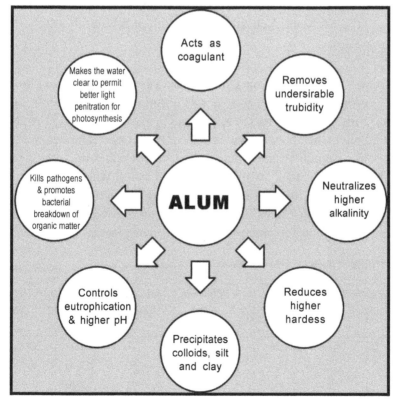

Fig. 10: Role of alum in reducing the environmental stress

is retarded by offering storage to the water, these impurities tend to settle down at the bottom of the tank offering such storage. This is the principle behind the sedimentation.

Fig. 11: Alum application, pH of eutrophic water (> 9.5)

Liming

Lime is frequently applied in aquaculture practices to improve water quality and acidity of sediments. There is a close relationship between the base saturation of bottom soils and total hardness of water. Total hardness of waters gets exceeded 20 mg/l when bottom mud is 80% or more bases saturated. Fertilization of ponds seldom produces good plankton growth if either the total hardness or total alkalinity is lesser than 50 mg/l. Some pond waters which contain more than 50 mg/l total hardness or total alkalinity, but are dark coloured because of the presence of humic substances, may also need lime to clean the water, permit better light penetration for photosynthesis and other factors (Fig. 12).

Table 17: Kinds of lime

Sl. No.	Chemical formula	Chemical name	Common name	Molecular weight	Relative neutralizing value (%)	Effect on water pH
1.	$CaCO_3$	Calcium carbonate	Limestone	100	100	Stabilizes at 7.5
2.	CaO	Calcium oxide	Quick lime	56	150-175	Enhances > 7.5
3.	$Ca(OH)_2$	Calcium hydroxide	Hydrated or slaked lime	74	120-135	Enhances >8.0

Application of lime dose

Lime and alum are two primary water purifying chemicals. If alkalinity or hardness is lesser than 50 mg/l, apply lime @ 5 mg/l; generally more than 10 mg/l may be counterproductive being as overdose. The correct dose can be ascertained by jar test procedure.

Estimation of lime requirement of ponds

Fertilization of pond seldom produces good plankton growth if either the total hardness or total alkalinity is lesser than 50 ppm. Response to fertilization is variable in ponds with 20-50 ppm total hardness and total alkalinity but waters above 50 ppm consistently produce phytoplankton growth after inorganic fertilization.

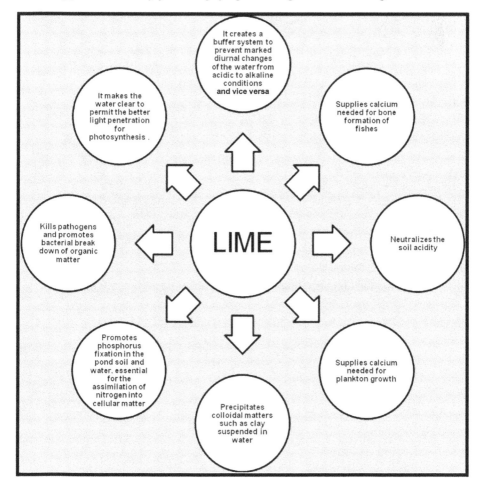

Fig. 12: Role of lime in reducing the environmental stress

Therefore, liming is indicated if either the total hardness or the total alkalinity of pond is below 50 ppm.

Lime application in pond

Lime stone (calcium carbonate, $CaCO_3$) under prolonged heating in calcining process gets converted into quick lime or (calcium oxide, CaO) which on hydrolysis gets changed into hydrated/slaked lime (calcium hydroxide, $Ca(OH)_2$.

$$CaCO_3 \text{ (Lime stone)} \xrightarrow[\text{PROCESS}]{\text{CALCINING}} CaO \text{ (quick lime)} + CO_2$$

$$CaO + H_2O \qquad\qquad Ca(OH)_2 \text{ (Hydrated or slaked lime)}.$$

Fig. 13: Lime application, pH of eutrophic water (< 9.5)

CHAPTER 9

CHLORINATION, DECHLORINATION, OZONATION, OXYGENATION AND AERATION

Introduction

Chlorination, dechlorination, ozonation, oxygenation and aeration are the practices to make the raw water free from pathogenic load which can not be conducive for hatchery and culture practices needed water quality criteria. Out of these practices, chlorination needs further its dechlorination of the treated water as its leaves behind the residual chlorine which at the level of more than 0.003 ppm (3 ug/l) may cause the mortality of reared aquatic animals.

Chlorination

Chlorine in its various forms is invariably and almost universally used for disinfecting public water supplies as well as for shrimp and fish hatcheries' requirements, and it is sometimes used to disinfect the production ponds before fish/prawn stocking. It is cheap, reliable, easy to handle, easily available and measurable; and above all, it is capable of providing residual disinfecting effects for longer period; thus providing complete protection against future recontamination of water in the concerned distribution systems. For its wide acceptance as a water disinfectant there are following reasons:

1. It is readily available in all the three matter states as gas, liquid and solid.

2. It is comparatively cheaper.

3. It is easy to apply due to relatively high solubility (7000 ppm) at 20^0C.

4. It leaves a residual.

5. It is highly toxic to most of the microorganisms.

6. Best oxidising power is at 7.2 to 7.6 pH.

Chlorine has some demerits as well, because it is a poisonous gas which requires careful handling and thus dechlorination of treated water is needed before releasing it into the hatchery, rearing and culture systems. The persons applying the chlorine should wear waterproof outer ware to protect their skin, an approved chlorine mask and goggles or a face shield for eye protection. Further, chlorine store house should have ventilator facility towards the floor side as well as, also along with ceiling side; as it is about two and half times heavier than air. Thus minimizing its harmful effects on the person while getting entered in the store house.

The common commercial chlorine compounds for disinfection are available in all the three states *viz.*, chlorine gas (Cl_2), sodium hypochlorite/sodium oxychloride (NaOCl), the liquid; and calcium hypochlorite/calcium oxychloride ($CaOCl_2$), the solid. Hypochlorites are comparatively less pure than chlorine gas, thus they are less dangerous. However, they have the major disadvantage that they get weak in their strength over time while in storage. Temperature, light and physical energy can break down hypochlorites before they are able to react with pathogens in water. Among the hypochlorites, sodium hypochlorite solution (NaOCl) is readily available in glass bottles, 5 litre and 10 litre plastic containers and it is generally known as bleach liquor or chlorine bleach. The quantities of hypochlorite required for chlorinating known volumes of water at the required levels vary in case of different samples. The exact quantities of sodium hypochlorite to be taken in case of samples of different concentrations to chlorinate known volumes of water and dosages of chlorine required in the treatment of water are given in the tables, 18 & 19 and hypchlorite can be added manually to the tank water. The water in aquahatchery practices should be treated for chlorination by bleach liquor only, not by bleaching powder as it can enhance the hardness of water which can act as a limiting factor for rearing organisms. For culture practices, the water in bulk quantity can be chlorinated with bleaching powder ($CaOCl_2$) which contains approximately 30% to 33% available chlorine. Bleaching powder is prepared by passing chlorine gas over dry slaked lime.

$$Ca(OH)_2 + Cl_2 = CaOCl_2 + H_2O$$

Chemical reactions

Chlorine reacts with water to form a strong acid (hydrochloric acid) and a weak acid (hypochlorous acid or HOCl).

$$Cl_2 + H_2O \leftrightarrow HOCl + H^+ + Cl^- \qquad ...1$$

Further hypochlorous acid (HOCl) dissociates and gives rise to hydrogen ion (H^+) and hypochlorite ion (OCl^-) as follows –

$$HOCl \leftrightarrow H^+ + OCl^- \qquad ...2$$

Thus, chlorination of water yields four chlorine species: chlorine, chloride (Cl^-), hypochlorous acid and hypochlorite. Among these species, chlorine, hypochlorous acid and hypochlorite are called as free chlorine residuals. The disinfecting power of chlorine and hypochlorous acid are about 100-times greater than that of hypochlorite.

Factors that influence effective disinfection with chlorine are as follows:

1. **pH:** When chlorine gets dissolved in water it forms the mixture of hypochlorous (HOCl) and hydrochloric (HCl) acids (reaction-1). This reaction is pH dependent. Further, hypochlorous acid dissociates to form chlorite (OCl^-), reaction-2. At pH levels between 4.0 and 6.0, chlorine exists predominately as HOCl. At pH levels above 8, hypochlorite ions (OCl^-) are predominant. However, hypochlorite ions exist almost exclusively at pH levels of 9 and above.

2. **Temperature:** Temperature affects the disinfection process, because water could be treated most efficiently at higher temperature.

3. **Turbidity:** The turbidity load in the available raw water is bound to enhance the chlorine demand. The turbidity level of water is reduced considerably through coagulation-sedimentation process by the time it reaches to filtration chamber. Excessive turbidity load creates more chlorine demand. When wastewater is filtered to a reduced turbidity of one unit or less, most of the bacteria are removed. Particulate matter may also change the chemical nature of the water when the disinfectant is added.

4. **Mixing/induction:** An efficient diffusion device to assure rapid initial mixing of chlorine and wastewater is essential. The most effective way of assuring rapid mixing and improving safety is to use the gas induction unit. This unit replaces the conventional diffuser, mixer, ejector, booster pump and long water solution line, which are generally used when the chlorinating agents are in the solid and liquid states.

5. **Contact time:** Contact time is an important factor which determines the reduction and removal of following causative factors as mentioned in the table 18. For effective disinfection it requires a minimum contact period of 2-3 minutes.

Table 18: Chlorine dosages for water and wastewater treatment

Sl. No.	Purpose of chlorination	Chlorine dosage (ppm)
1.	Disinfection with free residual	1.0 - 5.0
2.	Ammonia (NH_3-N) removal	10 x NH_3 - N content
3.	Hydrogen sulphide (H_2S) removal	1.22 x 5 content to free sulphur, 8.9 x 5 content to sulphate
4.	Iron (Fe) removal	0.64 x Fe content
5.	Manganese (Mn) removal	0.65 x Mn content
6.	Algal control	1.0-10.0
7.	Slime control	1.0 - 10.0

Dechlorination

Dechlorination is a practice used to reduce or remove the chlorine discharge levels. Chlorine must be removed from hatchery/culture water as it is toxic to rearing or culturable organisms if it exceeds more than 0.003 ppm or 3 ug/litre. The dechlorination may be carried out by adding certain chemicals to water or by simply aerating the water. These chemicals are called dechlorinating agents. The common dechlorinating agents are sodium thiosulphate, sodium bisulphate, sodium sulphite, activated carbon, potassium permanganate, suphur dioxide *etc.* and they help to reduce free and combined chlorine residuals. For instance free and combined chlorine residuals can be reduced by sulfur dioxide which gets dissolved in water rapidly, forming sulfuric acid as shown in the following reaction:

$$SO_2 + H_2O \rightarrow H_2SO_3$$

The sulfite radical formed in this solution reacts with free and combined chlorine as shown in the following equations:

$$H_2SO_3 + NH_3Cl + H_2O \rightarrow H_2SO_4$$
$$H_2SO_3 + NH_3Cl + H_2O \rightarrow NH_4HSO_4 + HCl$$

However, the most effective methods of chlorine removal are either air stripping the chlorinated water for overnight period or the treatment with sodium thiosulphate which reacts with chlorine (residual) as follows

$$2Na_2S_2O_3.5H_2O + Cl_2 = Na_2S_4O_6 + 2NaCl + 5H_2O$$

Molecular weight ratio =   496.2   : 70.9 ($2Na_2S_2O_3.5H_2O : Cl_2$)

Chlorine demand ratio =   7   : 1

Note: For the removal of 1 ppm of chlorine, 7 ppm of sodium thiosulphate is required.

Table 19: Sodium hypochlorite dosages for water and wastewater treatment

Quantity of water to be treated (litre)	Quantity of sodium hypochlorite (ml) required to chlorinate at the level of 10 ppm									
	1%	2%	3%	4%	5%	6%	7%	8%	9%	10%
1	1.00	0.50	0.32	0.25	0.20	0.16	0.14	0.12	0.11	0.10
2	2.00	1.00	0.64	0.50	0.40	0.32	0.28	0.24	0.22	0.20
3	3.00	1.50	0.96	0.75	0.60	0.48	0.42	0.36	0.33	0.30
4	4.00	2.00	1.28	1.00	0.80	0.64	0.56	0.48	0.44	0.40
5	5.00	2.50	1.60	1.25	1.00	0.80	0.70	0.60	0.55	0.50
6	6.00	3.00	1.92	1.50	1.20	0.96	0.84	0.72	0.66	0.60
7	7.00	3.50	2.24	1.75	1.40	1.12	0.98	0.84	0.77	0.70
8	8.00	4.00	2.56	2.00	1.60	1.28	1.12	0.96	0.88	0.80
9	9.00	4.50	2.88	2.25	1.80	1.44	1.26	1.08	0.99	0.90
10	10.00	5.00	3.20	2.50	2.00	1.60	1.40	1.20	1.10	1.00
20	20.00	10.00	6.40	5.00	4.00	3.20	2.80	2.50	2.23	2.00
30	30.00	15.00	9.60	7.50	6.00	4.80	4.20	3.75	3.30	3.00
40	40.00	20.00	12.80	10.00	8.00	6.40	5.60	5.00	4.40	4.00
50	50.00	25.00	16.00	12.50	10.00	8.00	7.00	6.25	5.50	5.00
60	60.00	30.00	19.20	15.00	12.00	9.60	8.40	7.50	6.60	6.00
70	70.00	35.00	22.40	17.50	14.00	11.20	9.80	8.70	7.70	7.00
80	80.00	40.00	25.60	20.00	16.00	12.80	11.20	10.00	8.80	8.00
90	90.00	45.00	28.80	22.50	18.00	14.40	12.25	11.25	9.90	9.00
100	100.00	50.00	32.00	25.00	20.00	16.00	14.00	12.50	11.00	10.00

Types of chlorination

Depending upon the quantum of chlorine added, or the stage at which it is added, or upon the results of chlorination, various technical terms in relation to chlorination are used.

Plain chlorination: This term is used to indicate that only the chlorine treatment and no other treatment has been given to the raw water. This technique may be used for treating relatively clear waters (with turbidity lesser than 20 to 30 ppm) obtained from lake or reservoirs *etc*. The usual quantity of chlorine required for plain chlorination is about 0.5 mg/l or more.

Pre-chlorination: Pre-chlorination is the process of applying chlorine to the water before filtration or rather before sedimentation–coagulation. It helps in improving coagulation controlling algae problems in basins, reducing odour and algal mat load (schmutzdecke) on filters. The normal doses required are as high as 5 to 10 mg/l.

Post-chlorination: The post-chlorination is adopted after filtration and before the water enters the distribution system. It is the application of chlorine when other treatments have been completed, but before the water reaches the distribution system. At this stage, chlorination is meant to kill pathogens and to provide a chlorine residual in the distribution system. It is nearly always part of the treatment process; either used in combination with pre-chlorination or used as the sole disinfection process. The dosage of chlorine should be such as to leave residual chlorine of about 0.1 to 0.2 mg/l, after a contact period of 20 min.

Double chlorination: The term double chlorination is used to indicate that the water has been chlorinated twice. The pre and post chlorination is generally used in double chlorination.

Break point chlorination: A method of chlorination in which chlorine is added to water until the chlorine demand has been completely satisfied (the breakpoint). Chlorine is added to pass the breakpoint for creating free chlorine residual.

Super chlorination: It is the term which indicates the addition of excessive amount of chlorine (*i.e.*, 5 to 15 mg/l) to the water. This may be required in some special cases of highly polluted waters or during epidemics of water borne diseases. It can deal with fishy, grassy or flowery odours and with iron and hydrogen sulfide but makes other problems worse and increases trihalomethane concentration.

Dechlorination: The term dechlorination means removing the chlorine from water. This is generally required when there is super chlorination or treated water is to be taken to culture and hatchery systems. The dechlorination process may either be carried out to such an extent that sufficient residual chlorine (0.1 to 0.2 mg/l) dose remains in municipal water supply; or otherwise if full chlorine has been completely removed, it is totally suitable for culture and hatchery practices.

Other terminologies associated with chlorination

Chlorine: A chemical element which has atomic number, 17 and atomic mass as 35.453. In its normal state, chlorine is a greenish yellow gas, but at -34°C it turns to a liquid. It is the eleventh most common element in the earth's crust and is widespread in nature; the most commonly used disinfectant for the deactivation of microorganisms and decolourization, and deodourization agent.

Chlorination: Chlorination is the destruction of waterborne pathogens through disinfection with various forms of chlorine (*e.g.*, sodium hypochlorite (bleach liquor), calcium hypochlorite (bleaching powder), chlorine gas, chloramines and chlorine dioxide). Thus, it is the application of chlorine.

Chloramines: Chemicals combining chlorine and nitrogen chloramines are formed by first adding chlorine gas or hypochlorite to water and then adding ammonia. Chloramines are a form of combined chlorine residual which can be used to disinfect water. They are weaker than chlorine gas, but are more stable, so they are often used as the disinfectant in the distribution lines of water treatment systems. They are monochloramine (NH_2Cl), dichloramine ($NHCl_2$) and trichloramine (NCl_3).

Chlorine demand: The total amount of chlorine which is used up in reactions with compounds in the water. A sufficient quantity of chlorine must be added to the water so that, after the chlorine demand is met, there is still some chlorine left to kill microorganisms in the water at the time of recontamination in distribution system.

Combined chlorine residual: A chlorine residual consisting of chlorine combined with nitrogen to form a chloramine. Combined chlorine residuals are sometimes used to disinfect water.

Chlorine residual: After treatment, a certain amount of chlorine will remain in the water. This amount is the chlorine residual. The chlorine residual must be maintained at a certain level throughout the distribution system in order to prevent further contamination. Chlorine testing in water can be done by colorimetric method through chloroscope taking 2-3 drops of orthotolidine as a reagent.

CT Value: Used as a measurement of the degree of pathogen inactivation due to chlorination.

The CT value is calculated as follows:

CT = (Chlorine residual, mg/l) (Contact time, minutes)

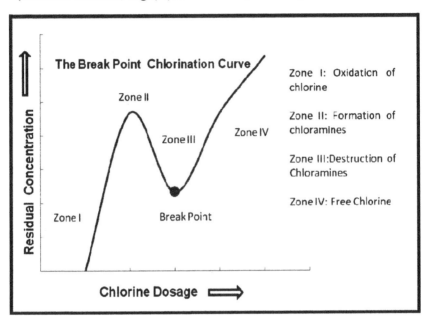

Fig. 14: The break point chlorination curve

Table 20: *Preparation of sodium hypochlorite (NaOCl) from the available material at site for hatchery practices

Sl. No.	Material required			
	Common name	Chemical name	Chemical formula	Quantity
1.	Bleaching powder	Calcium hypochlorite	$CaOCl_2$	71 g
2.	Washing soda	Sodium carbonate	Na_2CO_3	64 g
3.	Sea water	35 ppt water	H_2O(TDS: 35000 ppm)	1400 ml

Chemical equation

$$CaOCl_2 + Na_2CO_3 \longrightarrow CaCO_3 + NaOCl* + NaCl$$

*Application - as a prophylactic treatment measure in fresh and seawater

- as an oxidising agent for artemia cysts' decapsulation

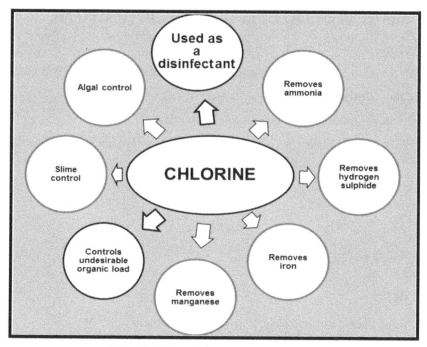

Fig. 15: Role of chlorine application

Safety warning

Never apply chlorine gas directly in waters through chlorine gas cylinder without the installation of induction unit; as direct chlorination through gas cylinder may cause explosion including full risk of occupational health hazard to the respiratory system of operator.

Conclusion

Chlorination is the best economical as well as efficient practice to make the water/wastewater free from pathogenic load. The most significant precautionary measure in this context is that, the chlorinated water should be dechlorinated prior to its use in a diversity of culture and hatchery systems.

Ozonation

In aqua production, a lot of organic wastes are introduced into the water which cannot be removed by normal filtering or sedimentation tanks. The ozonization of this water effectively not only eliminates these organic components by neutralization and oxidation but also kills bacteria in the circulating water. Ozone concentrations of 0.6 - 1.0 mg/l and contact time of 1-3 minutes are sufficient to kill most of the pathogens in aqua production systems. Safe permissible exposure of ozone to fish / prawn is 0.002 mg/l.

The merits of ozonization in aquaculture systems

- Prevents occurrence of water borne diseases
- Increases survival in packing and transport of fish and prawn seed
- Reduces turbidity created by organic matter
- Maintains optimum pH and dissolved oxygen
- Converts toxic nitrite into non-toxic nitrate through oxidation as nitrite is toxic to fishes at concentration as low as 0.5 ppm
- Neutralizes obnoxious gases like ammonia, hydrogen sulphide, methane and carbon dioxide *etc.*
- Favours increased growth of plankton and
- Reduces BOD and COD loads to trace level

Demerits

High installation cost and electric power requirement

Not available in all the three states like, solid, liquid and gas as in case of chlorine

To be generated at site by ozoniser machine

No residuals to prevent recontamination

Not easily available like chlorine

Can not be stored and packed like chlorine

Facts about ozone

It is tri-atomic allotrope of O_2, colourless gas with pungent odour.

It is one and half times as heavy as O_2 and thirty times more soluble in water than O_2.

It is very strong oxidant, 1.5 times stronger and faster disinfectant than chlorine.

The effective dose is 0.6- 1.0 mg/l for 1-3 min to make water safe from pathogens.

Safe permissible limit to cultivable organisms is 0.002 ppm (2 ug/l).

OXYGENATION AND AERATION

Dissolved oxygen concentration in aquaculture ponds varies depending on the various biological, physical and chemical processes, which add or remove dissolved oxygen from the water. Its concentration at any point of time depends on five major processes: (1) air-water gas transfer, (2) sediment oxygen uptake, (3) animal respiration, (4) plankton respiration and (5) photosynthesis. Although the budget appears quite simple because of the limited number of component processes, the rate of each process is affected by several physical and chemical factors and in fact, the dynamics of dissolved oxygen in ponds is quite complex.

Table 21: Comparative account on chlorination and ozonation

Sl. No.	Chlorination	Factor	Ozonation
1.	Low	Capital cost	High
2.	High	Operating cost	Low
3.	Yes	Raw material	No
4.	Poor or good	Control system	Excellent
5.	Low	Viricidal effect	High
6.	Yes	Toxicity of water	No
7.	Not so	Disinfection	Kill all microbes
8.	2-3	Contact time (min)	1-3
9.	7000	Max. Solubility (mg/l)	600
10.	0.003	Constant exposure of safe concentration (mg/l)	0.002
11.	1.0	Oxidative strength	1.5
12.	Taste and pH get changed	Effect on water	DO increase and more palatability
13.	Harmful at low conc.	Effect on human beings	Nil at low concentration

1. What is aeration?

The term aeration refers a mechanical process of bringing the water in intimate contact with air having an objective of supplying oxygen to water. Thus, aeration is a mechanical process while respiration is physiological one.

2. Importance of aeration

Dissolved free oxygen is essential for the survival and production of all cultivable species of fish and shell fish. Aeration is most effective means of increasing the DO concentration of culture water. The carrying capacity of an aquaculture system can be enhanced by applying suitable aeration device as the growth, feed efficiency and susceptibility to disease; and ultimately the production per unit area/volume and time largely depend upon DO availability. Therefore, it should not act as a limiting factor.

Types of aeration systems

Aerators for pond aquaculture are usually modifications of standard wastewater aeration equipments (Boyd, 1998). Aerators may be separated in two broad categories, splasher type and bubbler type (Boyd, 1998). Splasher aerators include three major types, *viz*., vertical pump aerators, pump sprayer aerators and paddle wheel aerators; while bubbler aerators are propeller-aspirator pump aerators and diffused-air aeration systems. All the basic types of mechanical aerators have been used in aquaculture but vertical pumps, paddle wheel and diffused air systems are most commonly used in pond aquaculture (Boyd, 1990, 1998). However, with aquaculture engineering point of view, there are four types of aeration systems, *i.e.,* gravity aerators, surface aerators, diffusers aerators and turbine aerators (Wheaton, 1977).

Causes of dissolved free oxygen depletion in aquatic systems

1) Lesser photosynthetic activity due to environmental factors

2) Meagre availability of chlorophyll bearing biomass, the phytoplankton

3) Higher organic load

4) Increase in temperature, salinity and TDS

5) Intensive stocking density

3.   Precautions

Lubricants, oil, diesel and grease should not reach to culture pond water.

Dissolved oxygen in pond is at the peak point in the afternoon, it declines during night and the lowest concentration occurs in the early morning hours (before sun rise), when aeration is more favourable. Therefore, aeration supply should not be stopped during early morning hours.

In the system, no dead points should be left over, *i.e.*, all parts of the system should have a reach of aeration.

The aeration supply column should be free from pond bottom to upper water surface thus to facilitate the oxygenation of all zones.

Maintain the favorable range of DO, *i.e.*, 5 ppm to 10 ppm.

4. Material required

Aerators are devices used to increase the concentration of dissolved free oxygen in the aquatic ecosystems by either supplying the air through air distribution system or by direct aeration. The following aerators can be required depending upon their suitable applicability.

A) **Gravity aerators :** Water falling through screen or a perforated plate Gravity driven paddle wheel and Plastic rings/wheels

B) **Surface aerators:**

Nozzles

Spray aerators and

Floating paddle wheel

C) **Diffuser aerators:**

Air stone diffuser

Venturi diffuser

U-tube diffuser

D) **Turbine aerator or propeller diffuser aerator**

5. Procedure

Atmosphere contains approximately 21% oxygen by volume. The rate of oxygen transfer into the pond water is an important factor that governs the intensity of aeration. The total atmospheric pressure is the sum of partial pressures of the gasses in the atmosphere, when the pressure of oxygen in water and air is equal; DO is said to be saturated. If the water is low in DO, *i.e.*, the DO is below saturation then the rate of oxygen transfer from the air is greater than what it would be if the water is nearest to oxygen transfer.

6. Observations

The following observations can be made when the aeration supply is continued in an aquatic system.

It enriches the pond water with oxygen and maintains the DO level in optimum range.

It removes the obnoxious gasses like ammonia, methane, hydrogen sulphide and carbon dioxide from the pond water.

It nullifies the activities of harmful and pathogenic bacteria.

It helps in decomposition of higher available organic load in the pond water by reducing the biochemical oxygen demand (BOD) and chemical oxygen demand (COD).

CHAPTER 10

PRINCIPLES AND APPLICATIONS OF AERATION IN AQUACULTURE

INTRODUCTION

Oxygen is the most important element for life. All life revolves around the availability of oxygen. In aquatic environments like lakes, fish ponds, aquaculture systems or aquariums, oxygen is the single most limiting factor for success. All aquatic life forms demand oxygen, from the fish down to the bacteria and any aquatic body has what is called an, "oxygen demand". Of all the essential elements and compounds in water, oxygen is also in the shortest supply. The air, we inhale is 21% oxygen. Water can hold the oxygen only about 8 parts per million or 0.0008% at $68^0F/20^0C$. It does not take long for oxygen to get depleted; then it is a job to make sure it is replenished continuously.

Water bodies get oxygen in two ways, 1) Photosynthesis by algae and plants, 2) Atmospheric oxygen is diffused into water at the pond surface. Photosynthesis can produce sufficient oxygen for low fish loads but it is inconsistent due to the fact that at night, algae and plants consume oxygen through the process of respiration. Most of the ponds and other water bodies with aquaculture intervention have higher than natural fish loads and thus require an additional oxygen demand to maintain a clean and healthy aquatic environment. To properly aerate aquatic bodies/systems, the available surface area is to be increased and improve circulation to enhance the diffusion of atmospheric oxygen into the water. Surface turbulence increases the available surface area in contact with the atmosphere and allows increased transfer of oxygen. Enhanced circulation evenly diffuses the oxygen to all zones within the pond. All techniques employed to aerate a body of water focus on increasing the surface turbulence and thus in turn enhancing the surface area for oxygen transfer. Most of these techniques also improve water and air circulation.

Aeration is the process of bringing water and air into close contact in order to remove dissolved gases, such as carbon dioxide, and to oxidize dissolved metals such as iron. It can also be used to remove volatile organic chemicals (VOC) in the water. Aeration is often the first major process at the water and wastewater treatment plants. During aeration, constituents are removed or modified before they can interfere with the treatment processes. Aeration and mixing are necessary components for aerobic biological wastewater treatment processes to maintain a healthy, stabilized biological population and thus, ensure optimal treatment of the wastewater assuming that there is adequate contact time and substrate available. For the design of wastewater treatment plants, aeration typically involves supply of both oxygen and mixing. Aeration is an important consumer of both energy and operational costs.

Principle of aeration: How does aeration remove or modify constituents?

The aeration process brings water and air into close contact by exposing drops or thin sheets of water to the air or by introducing air bubbles and letting them rise in the water column. For both procedures, the processes by which the aeration gets accomplished, the desired results are the same:

- Sweeping or scrubbing action caused by the turbulence of water and air mixing together.
- Oxidizing certain metals and gases. Undesirable gases enter the water either from the air above the water or as a by-product of some chemicals or biochemical reactions in the water.

The scrubbing process caused by the turbulence of aeration physically removes these gases from solution and allows them to escape into the surrounding air. Aeration can help to remove certain dissolved gases and minerals through oxidation, the chemical combination of oxygen from the air with certain undesirable metals in the water. Once oxidized, these chemicals fall out of solution and become suspended material in the water. The suspended material can then be removed by filtration. The efficiency of the aeration process depends almost entirely on the amount of surface contact between the air and water. This contact is controlled primarily by the size of the water drop or air bubble.

Importance of aeration

- Enrichment of the pond water dissolved oxygen.

- Improves growth rate of the culture animals.
- Reduces the stress on the aquatic animals.
- Removes the toxic or obnoxious gases in the pond water.
- Increases the organic decomposition at the pond bottom.
- Enhances the carrying capacity of the pond.
- Reduces the sludge build up in the pond bottom.
- Reduces the pathogenic activity.
- Breaks the stratification in the water column.

The reduced dissolved oxygen does not kill the fishes, but it has an impact on food intake, food utilization, normal metabolism and production yields. In extensive farming (low stocking density), the aeration is not needed. But in the case of intensive farming (high stocking density), the aeration is must. In shallow ponds, the oxygen is produced by macro vegetation and algae. In deeper ponds, the oxygen is produced by phytoplankton. Over manuring or fertilization of ponds leads to the over production of plankton which will inadvertently reduce the light penetration that leads to the oxygen depletion in deep waters. The following table elucidates the importance of aeration.

Table 22: Importance of aeration

Sl. No.	Before aeration	After aeration
1.	The excess growth of algal biomass in the pond water surface affects the light penetration.	Proper circulation of pond water improves the light penetration into the water column.
2.	The pond bottom does not receive the sufficient dissolved oxygen for the decomposition of organic matter which leads to the sludge formation.	Dissolved oxygen supply improves the organic decomposition and it reduces the formation of sludge in the pond bottom.
3.	The formation of sludge leads to the anaerobic condition in the pond bottom. So it does not support the aquatic animal life.	The supply of dissolved oxygen to the pond bottom supports the aquatic animal life.
4.	The presence of obnoxious gases like NH_3, H_2S and CO_2 reduces the aquatic productivity.	The addition of dissolved oxygen removes the obnoxious gases from water and enhances the pond productivity.
5.	Inadequate aeration increases the stress on culture animals and pathogens' activities.	Aeration reduces the stress and pathogenic activities.

[Table Contd.

Contd. Table]

Sl. No.	Before aeration	After aeration
6.	The middle layer only supports the aquatic animal life, (surface layer – algal blooms; bottom – anaerobic condition).	Entire water column supports the aquatic animal life.
7.	Poor aeration of pond water deteriorates the water quality.	It maintains the optimal water quality in the culture system.
8.	The absence of water circulation can lead to the formation of thermal layers in the water column.	It destratifies the formation of thermal layers in the water column.

Properties of dissolved oxygen

i.   The favourable range of dissolved oxygen concentration in aquaculture practices is 5-10 mgl^{-1}.

ii. The maximum DO solubility in water at -4^{0}C and normal atmospheric pressure is 16 ppm.

iii. The solubility of gases in water is inversely proportional to the availability of total dissolved solids, salinity and temperature.

iv. The equal volume of freshwater at the same temperature and pressure will hold more amount of dissolved gases than brackish and sea water.

v.   Maximum water density is at 4^{0}C temperature.

vi. Sea water normally contains 20% lesser dissolved oxygen than freshwater.

vii. The solubility of CO_2 in water can be 28 times more than oxygen and oxygen solubility in water is almost double than that of nitrogen present.

Chemical substances get affected by aeration

Aeration of water removes gases or oxidizes impurities, such as iron and manganese, so that they can be removed later in the treatment process. The constituents that are commonly affected by aeration are:

● Volatile organic chemicals, such as benzene, found in gasoline, or trichloroethylene, dichloroethylene and perchloroethylene, solvents which are used in dry-cleaning or industrial processes.

● Carbon dioxide

● Hydrogen sulfide (rotten-egg odour)

- Methane (flammable)
- Iron (will stain clothes and fixtures)
- Manganese (black stains)
- Various chemicals causing taste, odour and colour

Carbondioxide

Carbondioxide is a common gas produced by respiration. Apart from being naturally present in the air, it is produced by the combustion of fossil fuels. It is used by plants in the photosynthesis process. Surface waters have low carbondioxide content, generally in the range of 0 to 2 mg/l. Water from a deep lake or reservoir can have high carbondioxide content due to the respiration of microscopic animals and lack of abundant plant growth at the lake bottom. Concentration of carbondioxide varies widely in groundwater, but the levels are usually higher than in surface water. Water from a deep well normally contains lesser than 50 mg/l, but a shallow well can have a much higher level, up to 50 mg/l. Excessive amounts of carbondioxide above a range of 5 to 15 mg/l in raw water can cause three operating problems:

- It increases the acidity of the water, making it corrosive. Carbondioxide forms a "weak" acid, H_2CO_3 (carbonic acid).
- It tends to keep iron in solution, thus making iron removal more difficult.
- It reacts with lime added to soften water, causing an increase to the amount of lime needed for the softening reactions.

Most aerators can remove carbondioxide by the physical scrubbing or sweeping action caused by turbulence. At normal water temperature, aeration can reduce the carbondioxide content of the water to as little as almost nil.

Hydrogen sulfide (H_2S)

A poisonous gas, hydrogen sulfide can cause harmful problems in water treatment. Brief exposures lesser than 30 minutes to hydrogen sulfide can be fatal if the gas is breathed in concentrations as low as 0.03 percent by volume in the air. The Immediate Dangerous to Life and Health (IDLH) level for hydrogen sulfide is 300 ppm. Hydrogen sulfide occurs mainly in groundwater supplies. It may be caused by the action of iron or sulphur reducing bacteria in the well. The rotten-egg odour often noticed in well waters is caused by hydrogen sulfide, which in a water supply will disagreeably alter the taste of coffee, tea and ice. Hydrogen sulfide gas is corrosive to piping, tanks, water heaters and copper alloys that it contacts. Occasional disinfection of the well can reduce the bacteria producing the hydrogen sulfide. Serious operational problems occur when the water contains even small amounts of hydrogen sulfide:

- Disinfection of the water can become less effective because of the chlorine demand exerted by the hydrogen sulfide.
- There could be corrosion of the piping systems and the water tanks.

Aeration is the process of choice for the removal of hydrogen sulfide from the water. The turbulence from the aerator will easily displace the gas from the water. The designer of the system needs to consider how the gas is discharged from the aerator. If the gas gets accumulated directly above the water, the process will be slowed and corrosive conditions can be created.

Methane (CH_4)

Methane gas can be found in groundwater. It may be formed by the decomposition of organic matter. It can be found in water from aquifers that are near natural gas deposits. Methane is a colourless gas that is highly flammable and explosive. When mixed with water, methane will make the water taste like garlic. The gas is slightly soluble in water and therefore, it is easily removed by the aeration of water.

Iron and manganese (Fe & Mn)

Iron and manganese minerals are commonly found in soil and rock. Iron and manganese compounds can get dissolved into groundwater as it percolates through the soil and rock. Iron in the ferrous form and manganese in the manganous form are objectionable for several reasons. Water containing more than 0.3 mg/l of iron will cause yellow to reddish-brown stains of plumbing fixtures or almost anything that it contacts. If the concentration exceeds 1 mg/l, the taste of the water will be metallic and the water may be turbid. Manganese in water, even at levels as low as 0.1 mg/l, will cause blackish staining of fixtures and anything else it contacts. Manganese concentration levels that can cause problems are 0.1 mg/l and above. If the water contains both iron and manganese, staining could vary from dark brown to black. Typical consumer complaints are that laundry is stained and that the water is red or dirty. Water containing iron and manganese should not be aerated unless filtration is provided.

Taste and odour

Aeration is effective in removing only those tastes and odours that are caused by volatile materials, those that have a low boiling point and will vaporize very easily. Methane and hydrogen sulfide are examples of this type of material. Many taste and odour problems in surface water could be caused by oils and by-products that algae produce. Since oils are much less volatile than gases, aeration is partially effective in removing them.

Ceramic plate diffuser | Simple aerator

Paddle wheel aerator with 4 impellers | Diffuser aerator

Air pump | Rotary submersible air blower

Root air blower | Aerator blower

Fig. 16: Aeration devices

Dissolved oxygen

Oxygen is injected into water through aeration; this is, in most cases, beneficial. It increases the palatability of the water by removing the flat taste. The amount of oxygen that water can hold is dependent on the temperature of the water. The colder the water, the more oxygen the water can hold. However, water that contains excessive amounts of oxygen can become very corrosive. Excessive oxygen can cause additional problems in the treatment plant by causing air binding of filters.

Types of aerators

Aerators fall into two general categories. They either introduce air into the water or water into the air. The water-to-air method is designed to produce tiny drops of water that fall through the air. The air-to-water method creates small bubbles of air that are injected into the water stream. All aerators are designed to create a greater amount of contact area between the air and water to enhance the transfer of the gases.

CHAPTER 11

CONTROL OF pH, ALKALINITY AND HARDNESS

The pH is the potency of hydrogen ion concentration in water; originally it is a French language term, "pouvoir hydrogène". It is linked with the acidity and alkalinity of aquatic medium, 1-7 is the acidic range, exact 7 is the neutral nature and 7-14 is the alkaline condition. The enhancement of pH/alkalinity can be attained by the application of lime and it is described in detail in chapter No.8. The reduction of pH level is needed only when the water system is highly eutrophic and it can be achieved by the application of alum and hydrochloric acid. Their doses can be ascertained by jar test procedure. The total concentration of divalent metallic cations like, Ca^{++}, Mg^{++}, Fe^{++} and Sr^{++} in water is called as hardness, while on the other hand, combined strength of carbonate (CO_3^{--}), bicarbonates (HCO_3^-) and hydroxyl (OH^-) ions form water alkanity. At a given point of time, out of these three alkanity causing ions, any two can be present. The colourimetric test of pH can be done by universal indicator; having the composition as, 0.375 g thymol blue, 0.125 g methyl red in 100 ml of 70% alcohol (C_2H_5OH). After applying 2/3 drops of universal indicator in the water sample; green colour indicates neutral pH, green blue and blue as alkaline pH, while yellow and red as acidic pH.

Table 23: Classification of water on the basis of its nature and TDS concentration

Sl. No.	TDS (mg/l)	Type of water	Dominance
1.	Zero	Distilled/deionized/white water	Nil
2.	< 275	Mineralized	Minerals
3.	—	Acidic	Anions of iron, sulphur, aluminum
4.	—	Neutral	Balanced equilibrium

[Table Contd.

Contd. Table]

Sl. No.	TDS (mg/l)	Type of water	Dominance
5.	—	Alkaline	Cations
6.	0-50	Soft	Carbonate, bicarbonate
7.	50-150	Moderately hard	Carbonate
8.	150-300	Hard	Sulphate, carbonate
9.	> 300	Very hard	Sulphate, chloride
10.	> 500	saline	Sodium, chloride
11.	500 – 30,000	Brackish	Sodium, chloride
12.	30,000 –37,000	Sea/marine	Sodium, chloride
13.	> 37,000	Brine/metahaline	Sodium, chloride
14.	—	Wastewater (unfit for human consumption)	BOD/COD
15.	—	Effluent (industrial wastewater)	Industrial discharge
16.	—	Heavy water	Deuterium, a stable hydrogen isotope of double mass

Table 24: Effects of acidic and hard water on rearing organisms in aqua-hatchery

Sl. No.	Parameter	Effects on rearing organisms	Remarks
1.	Acidic water	Low pH (pH < 7.0) ✓ Increases the susceptibility of disease ✓ Decreases fecundity and egg fertility ✓ Corrosion of outer membrane of eggs ✓ Lesser survival of spawn and larvae	The favourable range of pH is 7.0 – 8.5 (fresh water) and 8.3 - 8.7 (brackish/seawater)
2.	Hard water	Hardening and bursting of eggs and in turn their mortality	Imbalanced relative differential gradient of ionic composition in the organisms and aquatic environment

Table 25: Water hardness and its control techniques

Acidic water		Hard water	
Parameter	Treatment	Parameter	Treatment
Acidity	- Application of sodium bicarbonate ($NaHCO_3$) @ 2 ppm for 1 ppm removal of acidity. - Rate of survival (post larvae), 42-45%	pH/ Hardness	- Application of sodium carbonate (Na_2CO_3) @ 5 ppm for 1 ppm removal of hardness - Use of zeolite ($Na_2 Al_2 SiO_2$, H_2O) (Ion Exchange Process) - Use of alum @ 5 ppm to bring down pH by 1 unit - Rate of survival (post larvae), 42-52%

The hardness in water is mainly caused by four dissolved compounds and these are as follows:

(1) Calcium carbonate/ bicarbonate

(2) Magnesium carbonate/ bicarbonate

(3) Calcium sulphate

(4) Magnesium sulphate

The presence of any of these compounds produces hardness. There are others, which are of lesser importance. Chlorides and nitrates of calcium and magnesium can also cause hardness but they occur generally in small amounts. Iron, manganese and aluminium compounds also cause hardness but as they are generally present in such small amounts, it is customary not to consider them in connection with hardness. The non-carbonate hardness, generally designated as permanent hardness is due to calcium and magnesium sulphates, chlorides and nitrates; while the carbonate and bicarbonate hardness is termed as temporary hardness. Hardness in water is expressed in terms of ppm or mg/l and sometimes milli equivalents per litre (meq/l). One meq/l of hardness producing ions is equal to 50 mg $CaCO_3$ (50 ppm) in one litre of water; while one degree hardness is equal to 14.25 ppm. Soft water has the dominance of carbonates and bicarbonates while hard waters are dominated by sulphate (SO_4^-) ions. The terms, soft and hard water are used when the levels of hardness are as follows:

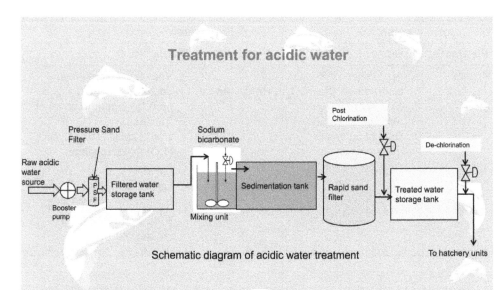

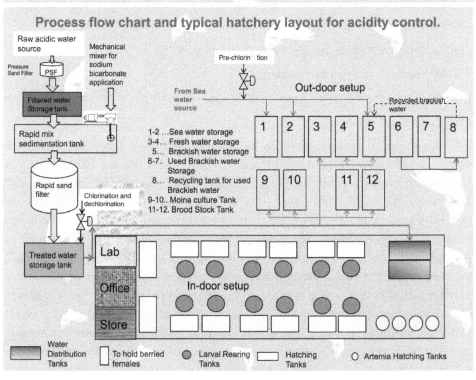

Fig. 17: Quality control of acidic water

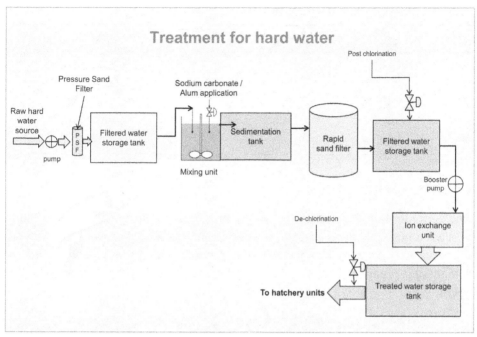

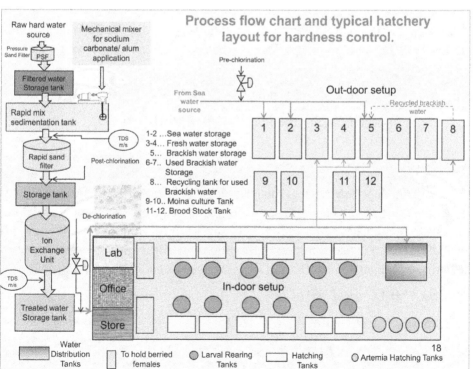

Fig. 18: Quality control of hard water

Table 26: Classification of hardness in water

Sl. No.	Level of hardness (ppm)	Type of water
1.	0 - 50	Soft
2.	50 - 150	Moderately hard
3.	150 - 300	Hard
4.	> 300	Very hard
5.	> 500	Saline

Control techniques

The concentration of hardness in culture practices should be more than 50 ppm to have a good plankton growth in pond water, on the other hand, in hatchery practices; it is better to be about 50 ppm to avoid scaling on the exoskeleton of soft rearing organisms. The methods of hardness removal are briefly stated as below.

Boiling

It removes the temporary hardness by expelling carbon dioxide and precipitating the insoluble calcium and magnesium carbonates. It is an expensive method to soften water on a large scale.

$$Ca(HCO_3)_2 + heating = CaCO_3 + H_2O + CO_2$$
$$Mg(HCO_3)_2 + heating = MgCO_3 + H_2O + CO_2$$

Addition of lime

Lime softening not only reduces total hardness but also accomplishes magnesium reduction. Lime absorbs the carbon dioxide and precipitates the insoluble calcium carbonate. In the Clark's method of softening water, one ounce (28.3495 gm) of quick lime (CaO) is added to every 700 gallons* (British units) of water for each degree (14.25 ppm) of hardness (Dh). One degree hardness = 14.25 ppm.

$$CaO + H_2O = Ca(OH)_2$$
$$Ca(OH)_2 + CO_2 = CaCO_3 \downarrow + H_2O$$
$$Ca(HCO_3)_2 + Ca(OH)_2 = 2CaCO_3 \downarrow + 2H_2O$$
$$Mg(HCO_3)_2 + 2Ca(OH)_2 = 2CaCO_3 \downarrow + Mg(OH)_2 \downarrow + H_2O$$

Addition of sodium carbonate

Sodium carbonate (Na_2CO_3) or soda ash removes both temporary and permanent hardness as shown below:

(i) $Na_2CO_3 + Ca(HCO_3)_2 \longrightarrow 2NaHCO_3 + CaCO_3$

(ii) $Na_2CO_3 + CaSO_4 \longrightarrow Na_2SO_4 + CaCO_3$

Base Exchange process

The most reliable method to control water hardness in aqua hatchery practices is the, 'Base Exchange Process'. As the name implies, ion exchange is a process in which undesirable ions are exchanged for more desirable ions. For the treatment of large water supplies this permutit process is used. Sodium permutit is a complex compound of sodium, aluminium and silica ($Na_2Al_2SiO_2$. H_2O). It has the property of exchanging the sodium cation for the calcium and magnesium ions in water. When hard water is passed through the permutit (Zeolite), the calcium and magnesium ions are entirely removed by Base Exchange and the sodium permutit is finally converted into calcium and magnesium permutit. By this process water can be softened up to zero hardness. After permutit has been used for some time, it loses its effectiveness but it may be regenerated by treating with concentrated solution of sodium chloride or brine (100 ppt - 150 ppt saline solution) and thus relinquish away the soluble calcium and magnesium chloride formed. This is called the service cycle. The regeneration is done at regular interval. This process removes both temporary and permanent hardness.

(A) Reactions during softening

(i) $Na_2Ze + CaSO_4 \longrightarrow CaZe + Na_2SO_4$

(ii) $Na_2Ze + MgSO_4 \longrightarrow MgZe + Na_2SO_4$

(B) Reactions during regeneration

(i) $CaZe + 2NaCl \longrightarrow Na_2Ze + CaCl_2$

(ii) $MgZe + 2NaCl \longrightarrow Na_2Ze + MgCl_2$

*One gallon = 4.546 liters (British unit); as practiced in India, one gallon =3.785 liters (U.S. unit)

* The dose of 2 ppm sodium bicarbonate ($NaHCO_3$) can reduce acidity by 1 ppm and 5 ppm of sodium carbonate (Na_2CO_3) can reduce hardness by 1ppm. The pH of acidic waters, if required to raise for hatchery operations, can be done by applying sodium bicarbonate without affecting the hardness through jar test procedure.

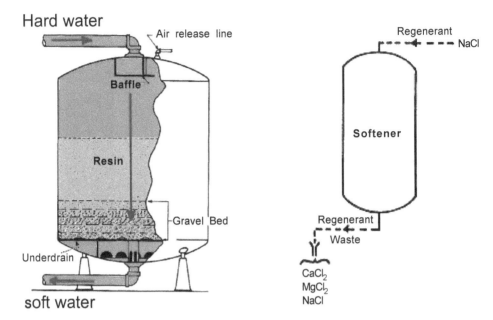

Fig. 19: Water softener

Quantification of resin

Resin quantity = Load (ppm as $CaCO_3$) x flow x time/Exchange capacity

For example:

Load = Hardness = 100 ppm as $CaCO_3$

Flow = 5 m³/hr

Time = service cycle = 12 hrs

Exchange capacity = 60 g as $CaCO_3$

Resin quantity = 100 ppm or mg/l x 5 m³ x 12 hrs/60 g

= 100 mg x 5 x 1000 litre x 12 hrs/60 x 1000 mg

= 100 x 5 x 12/60

= 100 x 60/60

= 100 litres

Lime soda softening process

It involves the application of lime $(CaOH)_2$ and soda ash (Na_2CO_3) to the hard water, which react with the salts of calcium and magnesium and in turn form insoluble precipitates of calcium carbonates $(CaCO_3)$ and magnesium hydroxide

$Mg(OH)_2$ Lime helps in removing carbonate hardness and substitutes the calcium salts for the magnesium salts. On the other hand, soda ash acts on the non-carbonate hardness of calcium salts.

$$Ca(HCO_3)_2 + Ca(OH)_2 = 2CaCO_3 \downarrow + 2H_2O$$

$$Mg(HCO_3)_2 + 2Ca(OH)_2 = 2CaCO_3 \downarrow + Mg(OH)_2 \downarrow + H_2O$$

$$MgSO_4 + Ca(OH)_2 = Mg(OH)_2 \downarrow + CaSO_4$$

$$CaSO_4 + Na_2CO_3 = CaCO_3 \downarrow + Na_2SO_4$$

$$MgCl_2 + Ca(OH)_2 = Mg(OH)_2 \downarrow + CaCl_2$$

$$CaCl_2 + Na_2CO_3 = CaCO_3 \downarrow + 2NaCl$$

Reverse osmosis and RO membrane

The RO membrane is the central system of reverse osmosis process, which is made of semipermeable material like cellulose acetate, aromatic polyamides or polysulfonated polysulfone. The membrane has pore diameter ranging from 0.5 nm to 1.5 nm. Reverse osmosis (RO) is the finest water purification available technology, which involves passage of water from the section having lower concentration of ionic substances to a higher concentration section when these two are separated by a semipermeable membrane. In this process, sufficient pressure is applied to force water to flow through the membrane in a reverse direction. Thus, purified water is produced by screening out dissolved solids and contaminants. Thus, improving colour, taste, odour and other properties of inlet/feed water. All dissolved solids and the particles with weight > 150-250 daltons can be removed/rejected by RO to the highest extent. The mechanical filters can remove suspended particles from water but RO can remove high quantum of dissolved contaminants – molecule by molecule from water. The RO is a cross flow or tangential flow filtration system with feed water flowing parallel to the membrane instead of perpendicular to it. Thus, the reject water sweeps rejected salts away from membrane surface ensuring not to clog the membrane pores. The installation of reverse osmosis system is more useful in hatchery practices to bring nonoptimal higher concentration of available source water hardness and total dissolved solids under optimal level.

CHAPTER 12

CONTROL OF AMMONIA, IRON AND HEAVY METALS

In culture and hatchery waters, ammonia is available in two forms, *i.e.,* ionized (NH_4^+-N) and unionized (NH_3-N). Unionized form is almost ten times more injurious than ionized one. Ammonia toxicity is directly proportional to the temperature and pH of aquatic medium. The safer limit for both, culture and hatchery practices is < 1.0 ppm (1000 ug/l) for ionized ammonia and < 0.1ppm (100 ug/l) for unionized ammonia at about 20^0c and pH value 7. One unit pH enhancement can increase ammonia toxicity by 10 times. If the available water at source is having the ammonia values more than these limits, the water needs ammonia removal. It can be accomplished by chlorination @ ammonia content×10. In culture practices, chlorination is to be done by bleaching powder $(CaOCl_2)$; on the other hand in hatchery practices by bleach liquor. As bleaching powder contains calcium ions which enhance the water hardness. The enhanced water hardness may cause the problems to younger ones' health in hatchery practices; therefore, the application of bleach liquor (NaOCl) is advisable, as it is not having hardness enhancing calcium ions in its chemical composition.

Iron content in natural waters is available in two forms, ferrous (Fe^{++}) and ferric (Fe^{+++}). Ferrous ions are available in ground waters, while ferric in surface waters. Its concentration > 0.3 ppm (300 µg/l) may cause the pathological response and in turn mortality of rearing and cultivable organisms; and concentration >1.0 ppm will cause mortality of culturable organisms, as at this stage, it tend to becomes highly toxic. Control of the iron for hatchery and culture waters can be done by application of chlorine dose @ iron content×0.64. The severe iron problem is faced in the available raw waters where either pyrites rocks are there particularly in the coastal belts or subsoil water of laterite belts. The metals, whose density is more than 6 g/ml, fall under the category of heavy metals. They can be removed by three ways,

1. Activated carbons (partial removal)

2. Surpentine mineral/chrysolite/antigonite ($3MgO.2SiO_2.2H_2O$)

3. Chelating agents (like EDTA).

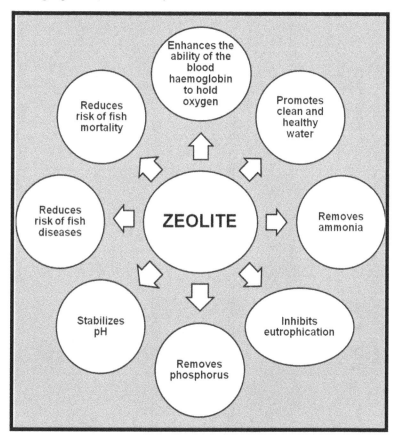

Fig. 20: Role of zeolite

Application of doses can be ascertained by jar test procedure

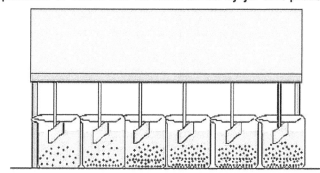

Fig. 21: Jar test apparatus for quantification of doses (Diagrammatic view)

Table 27: Comparative account on lime, alum, zeolite and chlorine

Lime	Alum	Zeolite	Chlorine
$CaCO_3$	$Al_2(SO_4)_3 \cdot 18H_2O$	$(Na_2[(AlO_2)_{12}(SiO_2)_{12}] \cdot 27H_2O)$	Cl_2
Neutralizes soil acidity	Excellent coagulant	Highest CEC	Best disinfectant
Precipitates colloidal matter such as clay suspended in water	Reduces turbidity	Removes ammonia, H_2S and toxic gases (1.5 ppm ammonia removal needs 1 ppm of zeolite)	Control and removal of Fe, Mn, NH_3, H_2S, slime, algal and organic load
Kills pathogens and promotes bacterial breakdown of organic matter	Controls eutrophication and higher pH	Stabilizes pH, removes phosphorous	Checks recontamination and oxidizes organic load

CEC: Cation exchange capacity.

Jar Testing

It is a well known pilot scale laboratory test for the selection, quantification and administration of chemicals in water and wastewater treatment practices. It helps operators to determine that the right amount treatment is being done coupled with efficiency of the treatment. Jar Test Apparatus is the six-place gang stirrer which can be utilized to simulate different processes of water and wastewater treatment. To ascertain the correct amount of dose, put the same and known water/wastewater quantum of the sample in the different jars. Prior to put the same volume of sample in jars, observe the initial value of parameter, which needs the dose application for chemical treatment. This nonoptimal observed value of concerned parameter is deviating how much from its optimal level is known now. Put the different fractions of the dose (solid or liquid) in jars, switch on to start apparatus operation; after 30 minutes operation, switch off and experimentally observe which fraction is giving desired result to achieve the optimal level. Select this dose fraction and quantify total amount of dose required as per the volume of water to be treated.

Fig. 22: Jar test apparatus for quantification of doses
(Photographic view)

CHAPTER 13

CONTROL OF BOD AND COD

Basically, BOD and COD are the units and test standards to quantify the aquatic pollution load. Biochemical oxygen demand (BOD) is the quantity of oxygen required by a definite volume of the water or wastewater for decomposing the organic matter present in it by microorganisms under aerobic conditions, per unit of time and temperature. For its determination, the dissolved oxygen content of the sample, with or without dilution, is measured before and after incubation at 20°C for 5 days period. Chemical oxygen demand (COD) is the amount of oxygen required to oxidize long term decomposable organic compounds like, lignin, cellulose, tannin, saw dust *etc.* in the presence of chemical oxidant like potassium dichromate ($K_2Cr_2O_7$) in the water or wastewater and/or a measure of the capacity of water or wastewater to consume oxygen during the decomposition of organic matter and the oxidation of inorganic chemicals like ammonia and nitrite. BOD needs decomposition and COD needs oxidation for their testing procedures. The rapid removal of BOD and COD can be made by aeration, ozonation and chlorination, while at the slower rate it requires more surface area, which can be done through oxidation ponds like, aerobic, anaerobic and facultative.

For faster BOD decomposition, application of urea (N): (NH_2-CO-NH_2) and ammonium phosphate (P): (NH_4)$_3PO_4$ can be made @ N= BOD/17 and P= BOD/100 of water or wastewater. On an average, 1 Kg fish weight needs 300 mg DO/h; and 1 ppm BOD load requires 2 ppm of DO, 0.5 ppm chlorine and 0.25 ppm of ozone for its decomposition/oxidation. The 1 ppm of COD load can be controlled by 2 ppm of 100% hydrogenperoxide (H_2O_2). The 95% of total BOD load is the amount of algal mass that can be produced from the wastewater. The per day per person BOD load generated is 0.1 kg (100 gm). If all the organic matter in water is biodegradable, the COD = BOD_{21}.

Note

As per the Gazette notification of Government of India, Ministry of Environment and Forest 1998, BOD testing can be done at 27^0C for 3 days period. In temperate climate BOD estimation at 20^0C for 5 days period is equivalent to 35^0C for 3 days period under tropical climatic conditions.

About 175 years back, when for the first time, the problem of aquatic pollution was felt due to sewage discharge in The River Thames in the city of London (Great Britain), the BOD parameter for its quantification was coined. The 20^oC temperature denotes the average temperature of the Great Britain and five days is the period of the water flow from its origin to getting discharged in the sea in Great Britain.

CHAPTER 14

PREPARATION OF ARTIFICIAL SEA WATER AND SALINITY MANIPULATION

The various water media which can be required while operating a successful prawn hatchery can be as follows:

Table 28: Different water media for larval rearing

Sl. No.	Water medium	Salinity range (ppt)
1.	Freshwater	0 to 0.5
2.	Brackish water	0.5 to 30
3.	Sea water	30 to 37
4.	Brine or Metahaline water	More than 37
5.	Artificial Seawater	Depending upon the desirable salinity

The medium of higher salinity can be brought down in the desirable limit as follows:

$$S_1 x V_1 = S_2 x V_2$$

Where:

S_1 = salinity of one medium (ppt)

V_1 = volume of one medium (litre)

S_2 = salinity of another medium (ppt)

V_2 = volume of another medium (litre)

Change of medium by applying brine

The salinity of one medium from lower limit to higher desirable limit can be raised as follows;

To increase the salinity 1 ppt, apply the salt or brine (check and calculate the density) @ one g per litre to the freshwater to be manipulated, *e.g.*, 10 ppt salinity increase in 100 liters of freshwater needs 10 g x 100 = 1000 g of salt or if brine of 100 litres of 4 ppt salinity then it needs additional 6 g of salt per litre (6 g x 100 = 600 g).

Artificial sea water

Artificial or synthetic sea water can be prepared by dissolving major, minor and trace salts on consecutive days to avoid precipitation of the salts with constant aeration as follows:

Table 29: Major, minor and trace salts required to prepare 1M ^{3} of 15 ppt artificial brackish water, experimentally successful and economically feasible, (Reddy et al., 1991)

Sl. No.	Name	Chemical formula	Quantity (g)
Major salt			
1.	Sodium chloride	$NaCl$	12000
2.	Magnesium chloride	$MgCl_2$	1520
3.	Sodium sulphate	Na_2SO_4	1214
4.	Calcium chloride	$CaCl_2$	352
5.	Potassium chloride	KCl	188
Minor salt			
1.	Sodium bicarbonate	$NaHCO_3$	80
2.	Potassium bromide	KBr	40
Trace salt			
1.	Boric acid	H_3BO_3	1
2.	EDTA	$C_{10}H_{14}N_2O_8Na_2.2H_2O$	1

Salinity and sea water composition

It is the quantification of saline nature property of water comprising the concentration of TDS. Salinity is measured by the dimensionless unit, parts per thousand and expressed by the symbols either ppt or 0/00. In modern oceanographic studies, it is also measured by 'practical salinity unit' (psu) and one ppt = 1 psu x 1.004715. By weight (mg/l), the six most abundant ions in sea water are chloride (Cl^-: 19000); sodium (Na^+: 10500); magnesium (Mg^{++}: 1350); sulphate (SO_4^- as S: 885); calcium (Ca^{++}: 400) and potassium (K^+: 380), which make up about 93% of all sea water ionic strength and 99% of all sea salts.

CHAPTER 15

SOIL - WATER AND THEIR INTERACTIONS

INTRODUCTION

Soil is the uppermost loose profile of the earth while sediment is the material that gets settled at the bottom of an aquatic body. The success and failure of any aquaculture production system mainly depends on the quality of soil and water. In reality, however, these two interact with each other along the plane of contact between the two phases. This interaction between liquid and solid phases is a very common phenomenon which is related to the difference in atomic structure of both phases on either side of the plane of contact. In the particular case of the plane of contact between the ionic solid phase and the aqueous liquid phase of the soil, the high dielectric constant of the soil solution is more significant. The polar water molecule possesses a strong tendency to cause dissociation of surface groups of the solid. On the other hand, the liquid layer adjacent to the solid phase is modified in comparison to the bulk solution both with regard to structure and ionic composition. In fact, the interaction between the solid and liquid phase gives rise to the formation of a 'surface phase' situated between solid and liquid. As ponds are built from soil thus many dissolved and suspended substances in water are derived and getting contacts with soil. Pond bottom soil is the store house for many substances that get accumulated in pond ecosystem and therefore chemical and biological processes occurring in the surface layer of pond soils influence water quality and aquaculture production.

The original pond bottom is usually made of terrestrial soil and when the pond is filled with water, the bottom becomes wet. A mixture of soil material and water can be called mud. Solids get settled from the water and cover the pond bottom with sediment and in older ponds, the sediment layer may be several centimeters thick. Thus, the term sediment is often used in referring to the solids at the bottom of a pond. Embankments that impound water and form ponds are made of soil, which also forms a barrier to seepage so that ponds will hold water. The ponds receive water from runoff and wells. Some ponds may be filled entirely by runoff, water pumped from streams or lakes or from estuaries. Surface water and ground water used to fill and maintain water levels in ponds acquire dissolved and suspended substances through contact with soil and other geologic materials. Further, contact and exchange between water and soil occur while water is retained in ponds.

Substances constantly settle from pond water at the pond bottom, for example suspended solids in surface water that enter ponds, particles of soil and organic matter that are eroded from the pond bottom and inside of levees by water currents and wave action, manure and uneaten food from management inputs and remains of plants and animals produced within the pond. Substances may also enter the solid phase of the soil from aquatic phase through ion exchange, adsorption and precipitation. For example K in the water can be exchanged for other cation of the soil, P can be adsorbed by soil; calcium carbonates may precipitate from solution and become a part of soil matrix.

Substances that enter the soil may be stored permanently, or they may be transformed to other substances by physical, chemical or biological means and lost from the pond ecosystem. To illustrate, phosphorus adsorbed by a pond soil can be buried in the sediment and lost form circulation within the pool of available phosphorus. Organic matter deposited on the pond bottom is usually decomposed to inorganic carbon and released to the water as carbon dioxide. Nitrogen compounds may be denitrified by pond soil microorganisms and get lost to the atmosphere as nitrogen gas.

Bacteria, fungi, algae, higher aquatic plants, small invertebrates and other organisms known as benthos live in sediment and on the bottom soil. Crustaceans and even some species of fish spend much of their time on the bottom and many fish species lay eggs in nests built in the bottom. Benthos serve as food for some aquaculture species; they also involve in gas exchange, primary and secondary productivity, decomposition and nutrient cycling. Substances stored in the pond can be released to the water through ion exchange, dissolution and decomposition. Release of inorganic ions or compounds from the soil into the water through exchange or dissolution only occurs until equilibrium is obtained between the solid and solution phases of the substances. The equilibrium concentration of a dissolved substance is an important consideration in pond management. The equilibrium concentration of a nutrient may be too low for optimal phytoplankton growth, or the equilibrium concentration of a heavy metal may be high enough to cause toxicity to aquatic animals. Microbial decomposition is extremely important because organic matter is oxidized to carbon dioxide and ammonia and other mineral nutrients are released. Thus, in decomposition, carbon, nitrogen and other elements are mineralized or recycled. Carbon dioxide and ammonia are highly soluble and quickly enter in the water. They may be used as nutrients, but if their concentrations are too high they can be toxic to aquatic animals.

Metabolic activities of microorganisms in pond soils are critical factors in pond dynamics. Microorganisms use molecular oxygen in oxidizing organic matter to carbon dioxide. During decomposition, soluble organic compounds are produced that may enter the pond water before being completely oxidized and contribute to the dissolved-organic-matter fraction in the water. Soils with large inputs of organic matter have high levels of microbial activity and the microbial community may use oxygen faster than it can penetrate the soil from the water above. This leads to anaerobic conditions and the development of a microbial flora that can use organic compounds or oxidized inorganic compounds instead of molecular oxygen as electron and hydrogen acceptors in metabolism. This process is often called anaerobic respiration. Denitrification, organic acid and alcohol production and formation of nitrite, hydrogen sulfide, ferrous iron, manganous and methane are the results of anaerobic respiration. Products of anaerobic respiration can enter the water and some are toxic to aquatic animals. At this stage, the organic carbon concentration of sediment may go high, > 2.75%, which is a response, the action indication time for pond desiltation.

Critical soil properties

Soils consist of weathered minerals and organic matter. They are a product of interactions among parent material, climate and biological activity. It is well known that soils differ from place to place on the earth's surface and beneath a given site the soil profile consists of horizontal layers that change in characteristics with depth below the land surface. Soils have been studied in great detail and desirable soil characteristics for agriculture and engineering purposes have been identified and classified. Much less is known about desirable characteristics of soils for aquaculture ponds, but there is enough information available to identify several critical factors.

The most active fraction of the soil is clay particles, because of their electrical charge, large surface area, retention of organic matter, their biological availability and high chemical reactivity. The pore spaces among the mineral fragments and organic matter in soil are filled with air and water. In flooded soils, the air is completely displaced by water. This greatly impedes the movement of oxygen into the pond soil, because the oxygen must move by diffusion or be carried along with water that seeps through the soil. Coarse-textured submerged soils are normally better oxygenated than fine-textured ones. Pond soils are often fine-textured because they usually have at least 20-30% clay content to limit seepage and they usually have a higher percentage of organic matter than terrestrial soils in the surrounding area.

Table 30: Selected soil properties that influence aquaculture pond management

Sl. No.	Property	Process affected in pond
1.	Particle size and texture	Erosion, sedimentation, embankment stability, seepage, suitability of bottom habitat
2.	pH and activity	Nutrient availability, microbial activity, benthic productivity, hydrogen ion toxicity
3.	Organic matter	Embankment stability, oxygen demand, nutrient supply, suitability of bottom habitat
4.	Nitrogen concentration and C:N ratio	Decomposition of organic matter, nutrient availability
5.	Redox potential	Toxin production, mineral solubility
6.	Sediment depth	Volume reduction, bottom habitat suitability
7.	Soil colour	Ionic composition of subsoil water
8.	Nutrient concentration	Nutrient availability and productivity

Normally pond soils are highly reactive, have a high oxygen demand and tend to become anaerobic. In addition to texture and organic matter content, specific chemical compounds in soil may have a pronounced effect on aquaculture. Soils that have been highly weathered and contain appreciable quantities of aluminium oxides and hydroxides are acidic, and water in contact with acidic soils have low total alkalinity concentrations and are poorly buffered against pH change. The presence of iron pyrite in aerobic pond soil can result in extreme acidity because sulfuric acid is produced by pyrite oxidation. Free calcium carbonate in soil leads to high concentrations of total alkalinity and total hardness in overlaying water. The concentrations and proportions of major ions in surface water are governed by the kind of rocks and soil with which water has been in contact and the amount of rainfall relative to evaporation. In areas with easily weathered rocks or fertile soils, surface waters are usually more mineralized than in areas with resistant rocks or infertile soils. Surface waters of arid climates are more mineralized than those in humid climates. Within a given climatic region, ponds built on different kinds of soils will have different levels of mineralization and different proportions of major ions.

The four most important soil features for aquaculture production are texture, organic matter content, pH and presence or absence of particular soluble compounds that may be beneficial or harmful to water quality. When soils in ponds are flooded, the most marked change in their composition is loss of air from the pore space and the gradual accumulation of organic matter.

Table 31: Favourable range of soil parameters for aquaculture

Sl. No.	Parameter	Favourable range
1.	Nature	Sandy-clay-loams
2.	Colour	Blackish/brownish
3.	pH	6.0 to 8.0
4.	Water retention capacity	40% and above
5.	Sand	40% (particle size 2.0 – 0.02 mm)
6.	Silt	30% (particle size 0.02 – 0.002 mm)
7.	Clay	30% (particle size < 0.002 mm/2 micron)
8.	Total nitrogen	> 50 mg/100 g of soil
9.	Phosphate	> 6 mg/100 g of soil
10.	Potassium	> 25 mg/100 g of soil
11.	Organic carbon	0.5 to 2.75%
12.	Electrical conductivity	< 16 millimohs/cm

Reactions and processes

The following ones are important points to understand pond soil-water interactions; (1) most of the reactions and processes occur continuously and simultaneously; (2) two or more reactions may be for the same reactant; (3) different processes and reactions that are interdependent may occur at different rates; and (4) it is difficult to study more than one or two reactions or processes at a time, and frequently the necessary components can only be isolated and studied in laboratory systems. Most of the reactions and processes are discussed here individually, but interrelationship is stressed. To understand the individual reactions and processes, try to develop an overview of the pond soil –water system as a whole. Such a view is essential in assessing the influence of pond soils in practical pond management situations.

Pond soil management

Techniques available for overcoming limitations of site soil properties in pond construction and seepage control are discussed. After a pond has been constructed, a few techniques are widely used for mitigating deficiencies in pond soil conditions. Fertilizer may be periodically applied to replace phosphorus removed by bottom soil. Ponds are limed to increase the base content and pH of bottom soils so that pond water will have adequate total alkalinity and total hardness for aquatic productivity. Aeration and water circulation devices can be used to improve dissolved oxygen concentration in the water and at the pond bottom. Sedimentation basins can be used to remove suspended and undissolved particles from raw water before it is taken to ponds. Between crops, pond bottoms are often dried and tilled to improve aeration and to accelerate the decomposition of organic matter.

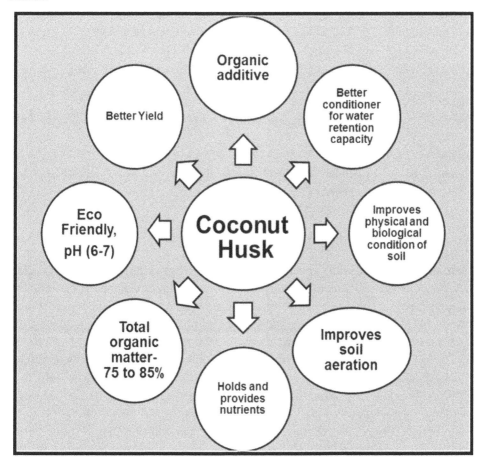

Fig. 23: Pond soil management through coconut husk

Table 32: Selected physical, chemical and biological reactions that occur in pond soil or at soil-water interface

Phenomenon	Pond soil or at soil-water interface (reaction example)
Dissolution	$CaCO_3 + CO_2 + H_2O = Ca^2 + 2HCO_3^-$
Precipitation	$Al^3 + H_2PO_4^- + 2H_2O = Al(OH_2)H_2PO_4 + 2H^+$
Hydrolysis	$Al^3 + 3H_2O = Al(OH)_3 + 3H^+$
Neutralization	$HCO_3^- + H^+ = H_2O + CO_2$
Oxidation	$NH_4^+ + 2O_2 = NO_3^- + 2H^+ + H_2O$
Reduction	$SO_4^{2-} + 4H_2 = S^{2-} + 4H_2O$
Complex formation	$Cu^{2+} + CO_3^{2-} = CuCO_3$
Adsorption	Adsorption of phosphorus on soil colloids
Cation exchange	K (soil) = K+ (water)
Hydration process	$Al_2O_3 + 3H_2O = Al_2O_3.3H_2O$
Dynamics	Not static but continuously changing
Sedimentation	Soil particles in runoff get settled at pond bottom
Decomposition	Microorganisms break down soil organic matter: $CH_2O + O_2$
Photosynthesis	Algae produce organic matter and release oxygen: $6CO_2 + 6H_2O = C_6H_{12}O_6 + 6O_2$
Diffusion	Oxygen diffuses into bottom soil from water column
Seepage	Water carrying dissolved substances seeps downward into pond soil
Erosion	Water waves in pond erode the bottom soil
Suspension	Particulate matter eroded from the bottom is suspended in pond water

Mechanisms of clay-water interactions

There are several possible mechanisms by which water may interact with clay surfaces. These mechanisms of interactions may operate separately or unitedly. All discerned is their net effect on the water. Hydrogen bonds are normally formed between O——H groups and oxygen atoms and there is a tendency for water molecules to be hydrogen bonded in a tetrahedral arrangement. The surface of clay minerals is made up of either oxygen atoms or hydroxyl groups arranged in a hexagonal pattern which can coincide at points with a similar pattern in a hydrogen-bonded water structure. Further, the crystal lattice of most clay minerals contains excess electrons which arise from the isomorphism substitution of cations in the lattice. The covalence may occur in hydrogen bond formation if one of the systems involved is capable of having its lone pair electrons distorted by the proton or positive element of the other. Such distortion is conducive to the formation of additional hydrogen bonds in a cooperative manner.

Now, the lone-pair electrons of the oxygen atoms in the surface of a clay mineral should be easily distorted because of the excess electrons in the lattice. Therefore, it is reasonable to believe that water molecules adjacent to a clay mineral surface are bonded to the oxygen atoms of the surface by covalent hydrogen bonds. The existence of the covalent bonds should alter the electron distribution in these molecules and make it easier for them to form additional covalent bonds with other molecules in the same and next layer. Those in the next layer, in turn, may be expected to form hydrogen bonds of partially covalent character with their neighbors and so on. The bonded water molecules should be arranged in a tetrahedral fashion because of the directional properties of the bonds. However, the degree of covalence in the bonds should decrease with distance from the surface and for this reason; the tetrahedral arrangement should become less rigid in the same direction. Thus, it is possible for a tetrahedral structure of water molecules to be attached to and propagated, with decreasing rigidity, away from the oxygen surface of a clay mineral.

It is not unlikely that a hydrogen-bonded water structure builds up also on the hydroxyl surface of a clay mineral. Here the excess of electrons in the mineral lattice should help to screen the protons of the hydroxyl groups and render them less electropositive. Consequently, the lone-pair electrons of the oxygen atoms in the bonded water molecules should experience little distortion and the degree of covalence in the hydrogen bonds should be slight. For this reason, the water structure on a hydroxyl surface may be expected to be less stable than that on an oxygen surface. But, the balance between order and disorder in water is delicate. Therefore, even the hydroxyl surface, by fixing the positions of a layer of bonded molecules, should tip the balance in favour of order for considerable distances.

Probably, the water structure that develops on either mineral surface is not that of ice. One reason is that the exchangeable cations would disrupt the water structure. Another is that the surface atoms of the mineral may not coincide exactly with protons or oxygen of the ice lattice so that the latter would be distorted (Mathieson and Walker, 1954). Or surface irregularities, which must exist, may produce distortions in the ice lattice. Further, there may be other hydrogen-bonded structures with tetrahedral coordination which are more stable in such an environment (Hendricks and Jefferson, 1938).

In laboratory studies, soil samples from three of the current CRSP research sites were treated with deionized water and/or suspensions of $CaCO_3$ in deionized water to evaluate soil-water interactions with respect to alkalinity in the water column. Samples were obtained from Butare (Rwanda), the Freshwater

Aquaculture Center of the Philippines, (FAC) and El Carao (Honduras). All samples from the Butare station had very low initial pH values (4.5 to 4.9) and base saturation percentage (18 to 48), whereas samples from El Carao and FAC were alkaline (pH 7.9 and 7.7 respectively and initial base saturation 100%). Samples from El Carao and the FAC were therefore treated only with deionized water. After 24 h of treatment (with agitation to promote complete reaction), the mean total alkalinity values (three replications) of the supernatant solutions for these soils were 62 and 32 mg $CaCO_3$/L respectively. The level of calcium and sodium in these soil samples were high relative to those of other soils tested, suggesting that in addition to calcium carbonate, sodium carbonate may have been present in quantities sufficient to release considerable alkalinity into solution. In addition, the exchangeable magnesium present in the FAC sample was higher than most other samples, whereas the amount of exchangeable potassium present in the sample from El Carao was unusually high. In contrast, soil samples from the Butare site treated only with deionized water had insignificant amounts of alkalinity (all lesser than 2.5 mg $CaCO_3$/L,); these samples all required treatment with concentrations of at least 0.75 millimolar $CaCO_3$ to reach alkalinity of 20 mg $CaCO_3$/L or greater.

When ponds are filled with water from surface runoff, streams or well water, the inflowing water will contain nutrients. In the pond, there will be an exchange of nutrients between soil and water until equilibrium is established. In the acidic soils, typical of humid regions, concentrations of inorganic carbon (carbon-dioxide and bicarbonate) have often been reduced to very low levels by leaching and concentrations in surface runoff are correspondingly low (Boyd, 1995). As a result, equilibrium concentrations of inorganic carbon in ponds in these areas are often too low to permit high rates of primary productivity. In semiarid areas, on the other hand, many soils contain calcium carbonate and are alkaline, and surface waters typically contain more bicarbonate, so equilibrium concentrations of inorganic carbon in ponds in these regions are often higher than those in humid areas. Most pond waters, regardless of pH and alkalinity, will not contain enough dissolved inorganic nitrogen (nitrate and total ammonia nitrogen) or phosphorus to support adequate primary productivity for high levels of fish production. Acidic ponds should be limed to elevate pH and total alkalinity and provide more inorganic carbon for phytoplankton. Chemical fertilizers and manures can be applied to ponds to enhance the availability of nitrogen and phosphorus for aquatic productivity. Nitrogen fixation is an abundant source of nitrogen in many ponds, and nitrogen fertilization is often not necessary. Bottom soils strongly adsorb phosphorus and most ponds require phosphorus fertilization. Soils with a low pH and high concentrations of iron and aluminium oxides are especially adsorptive of phosphorus, which is quickly lost from solution in ponds where waters have a high pH and a large calcium concentration (Boyd, 1990).

Banerjea (1967) studied the relationships between pond soil characteristics and fish production in many ponds in India. He stated that available nitrogen concentrations should exceed 250 ppm and available phosphorus concentrations should be above 60 ppm to support good fish production. Similar estimates have not been made for other regions, but almost all studies have shown that phosphorus applications will enhance phytoplankton productivity and in turn increase fish production, regardless of soil phosphorus concentrations. Nitrogen fertilization alone usually will not enhance fish production, even in ponds with low soil nitrogen concentrations. However, in some cases, nitrogen plus phosphorus fertilization will provide greater fish production than phosphorus fertilization alone (Boyd, 1990).

Acid sulphate soils

Acid sulphate soils or cat clay soils from coastal mangroves contain high levels of pyrites, the iron sulphate mineral (FeS_2 – 1.6%); thus it is a problem to brackish water aquaculture. As they get drained and exposed to the air, oxidation results and sulphuric acid is formed. It reduces the pH of the water when pond is filled. In ponds, the problems with acid sulphate soils usually originate in pond dykes. Pond bottoms are usually flooded and anaerobic, which prevent formation of sulphuric acid. However, sulphuric acid is formed when dykes get dried and it enters in ponds with runoff water after rains. Acidity on dykes can be controlled by liming @ 0.5 – 1.0 kg/m^2. The dykes can also be covered with the acid resistant perennial grass species. A procedure for rapid reclamation of ponds with acid sulphate soils involves drying and filling of the soil to oxidize pyrite, filling the ponds with water and holding till water pH drops to below 4.0 and draining the ponds, repeating the procedure, until the pH gets stabilized at a pH value 6.0 or above and then liming the pond with 500 kg of $CaCO_3$/ha.

$$2FeS_2 + 7O_2 + 2H_2O = 2FeSO_4 + 2H_2SO_4$$

Nutrient status of the soils

Nitrogen, phosphorus and potassium are the three major nutrients required by phytoplankton. Inorganic fertilizers can be applied to provide these nutrients. The appropriate dose depends on the amount of individual nutrients present in the soil in the available form. In general, relatively small amounts of potassium are needed in fish ponds. The single most critical nutrient for the maintenance of pond productivity is the available phosphorus content of pond soil and water. Pond soil with 30 ppm, 30 – 60 ppm, 60 – 120 ppm and more than 120 ppm available phosphorus (P_2O_5) are considered to have poor, average, good and high productivity respectively. Ponds with lesser than 250 ppm available soil nitrogen are considered to have low productivity while its availability of more than 250 ppm and >500 ppm are considered to be medium and highly productive respectively. Productivity and physico-chemical properties of soil can be improved with the application of organic manures and inorganic fertilizers, containing sufficient quantities of nitrogen and phosphorus.

Organic carbon content

It acts as the source of energy for bacteria and other microbes that release nutrients through various biochemical processes. Pond soils with lesser than 0.5% organic carbon is considered unproductive while those in the range of 0.5 – 1.5% and 1.5 – 2.75% are considered medium and high productive respectively. More than 2.75% organic carbon content can not be suitable for fish production. It can lead to excessive bloom of microbes, oxygen depletion and anaerobic condition in pond bottom sediment.

Carbon to nitrogen ratio

The C : N ratio of soil influences the activity of soil microbes to a great extent. This in turn affects the rate of release of nutrients from decomposing organic matter. Soil C: N ratio in between 10:1 and 15:1 is considered favourable for aquaculture.

Bottom soil oxidation

Dissolved oxygen can not move rapidly into water-saturated soil and pond soils become anaerobic below a depth of few millimeters. Aeration and water circulation are beneficial in improving bottom soil oxygenation, but the surface layer of soil may still become anaerobic in intensive fish culture ponds. When the redox potential is low (<100 mv) at the soil surface (anaerobic conditions), hydrogen sulphide and other toxic microbial metabolites diffuse into the pond water. Sodium nitrate ($NaNO_3$) can serve as source of oxygen for microbes in poorly oxygenated environments in which the redox-potential will not drop low enough for the formation of H_2S and the other toxic metabolites.

Desiltation/drying pond bottom

When pond bottom has high humic contents (organic carbon >2.75%), pond needs desiltation and drying up of its bottom surface area.

Bottom soil treatment

The following are the methods of bottom soil treatment; field experiment observations.

1. **Nitrogen application:** Ponds were treated with chicken litter at 750 kg dry weight per hectare and urea was added to determine if additional N_2 would enhance the degradation of chicken litter and decrease the rate of organic matter accumulation in the soil. There was an increase in soil carbon in the 0-5 cm soil layer in the ponds treated with chicken litter. Soil carbon could not get changed in the 5-15 cm soil layer. Thus, it appears that N_2 fertilization did not stimulate microbial activity and increase the decomposition of organic matter in pond bottom soils. Organic soils which have a much larger C: N ratio, envisages that N_2 application will stimulate decomposition.

2. **Aeration:** Aerated ponds have soil carbon increase in the 0-5 cm layer and beyond that no significant increase was measured in the culture ponds. This indicates that aeration to supply dissolved oxygen at the pond bottom can accelerate organic decomposition in bottom soils.

3. **Draining:** Experiments have shown that weight of the solids that might settle were lost during a single draining. When ponds were drained, the carbon concentrations in the 0-5 cm layer got declined, but had no influence on soil carbon in the 5-15 cm layer. The loss of soil carbon during draining may be considered as good effect on the pond environment, but the suspended soil particles in pond effluent represent a sediment load to receiving bodies of water.

4. **Drying:** After draining, the soil was allowed to dry for 5 weeks. When the soil dried, deep cracks got developed. Drying the soil removed free water from pore spaces and allowed air to enter the soil. The cracks also enhanced aeration. Soil carbon in the 0-5 cm layer decreased from 1.45 to 1.44% during 5 weeks of drying, the corresponding decrease of soil carbon in the 5-15 cm layer reached from 1.44 to 1.23%.

5. **pH adjustment:** The optimum pH for organic matter decomposition was 7.5-8.0 but pH values between 7 and 8.5 were all associated with much greater rates of soil respiration than lower and higher soil pH. Liming acidic soil to raise the pH above 7 appreciably increased soil respiration; a treatment rate of 1000-2000 kg/ha is sufficient for most of the soils. Agricultural limestone

is the best material to use in most cases, because even when applied @ 2000 kg/ha, its use does not increase soil pH above 7.5. When $Ca(OH)_2$ was applied @ 2000 kg/ha, there was an initial increase in the soil pH above 10. High pH caused an inhibition of soil respiration, but the inhibition was temporary. The $Ca(OH)_2$ or CaO are useful for killing fish pathogens, which may lie dormant in pond sediments.

6. **Duration of fallow period:**   A fallow period of 2-4 weeks is probably adequate for most soils. Those soils that dry quickly would need a shorter fallow period.

Table 33: Common liming material used for promotion of water hardness and soil pH

Sl. No.	Material	Common name	Chemical formula/ Molecular weight	Relative neutralizing value (%)	Rate of application based on pH of soil (Kg/ha)		
					pH 4.5-5.5	pH 5.5-6.5	pH 6.5-7.5
1.	Calcium carbonate	Lime stone	$CaCO_3$/100	100	2000	1000	500
2.	Calcium hydroxide	Hydrated/ slaked lime	$Ca(OH)_2$/74	120 - 135	1450	725	380
3.	Calcium oxide	Burned/ quick lime	CaO/56	150 - 175	1100	550	225

Conclusion

The pond soil is beneficial to the pond ecosystem as a basin to hold water, a storehouse of various chemical substances, a habitat for plants and animals and a nutrient recycling centre. It can also exert a large oxygen demand, may become anaerobic and be a source of toxic dissolved substances. Research and practical observations clearly reveal that pond soils influence water quality and production. In extensive aquaculture production, yields are better in areas where soils are fertile than in places where soils are infertile. Fertilized ponds with acidic soils typically have lower fish production than fertilized ponds with near-neutral or slightly alkaline soils. Ponds with higher concentrations of soil organic matter (>4.74%) develop anaerobic zones in their bottom and toxic metabolites from microbial activities retard growth or even cause mortality of aquaculture crops. The pH and concentrations of carbon, nitrogen and phosphorus in soils affect the potential of ponds for aquaculture production. Soils with pH below 4.8 frequently cause toxic levels of aluminium and manganese in turn immobilise the nutrients.

* Soil organic matter % = Actual carbon % x 1.724.

CHAPTER 16

INTEGRAL CALCULATION OF DOSES

INTRODUCTION

To make aquatic environment conducive for improved growth, health and survival in the hatching, rearing and culture of aquatic organisms, the quantification and administration of dosages are equally important. The effects of exposure for treatment depend on the dose, duration and the way of exposure. The common practice is to dilute solid or liquid chemicals in water and prepare the stock solution, working solution and treatment dose. In this context, some of the important examples are given for the calculation and application of solid and liquid dosages.

Example 1: Application of solid dose

1 ppm = 1mg/l

    = 10 kg/ha

LHS		RHS
1ppm	=	1 mg/l
1/10,00,000	=	1 mg/kg
	=	1mg/1000 gm
	=	1mg/1000 × 1000 mg
	=	1mg/10, 00,000 mg
	=	1/10, 00,000
Therefore, LHS	=	RHS

Example 2: Application of liquid dose

1 ppm = 1ml/ m^3

    = 10 liters/ha (average depth one meter)

LHS		RHS
1ppm	$=$	$1\text{ml}/1\text{ m}^3$
1/10,00,000	$=$	1 ml/1000 L
	$=$	$1\text{ml}/1000 \times 1000$ ml
	$=$	1ml/10, 00,000 ml
	$=$	1/10, 00,000
Therefore, LHS	$=$	RHS

Example 3: Salinity manipulation

1 ppt $= 1\text{g}/$ l

LHS		RHS
1ppt	$=$	1g/ l
1/1000	$=$	1 g/ 1 kg
	$=$	1g/ 1000g
	$=$	1/1000
Therefore, LHS	$=$	RHS

Example 4: Salinity reduction

$$S_1 V_1 = S_2 V_2 \text{ (dilution formula)}$$

Where, $S_1 V_1$: salinity and volume of one solution

$S_2 V_2$: salinity and volume of another solution

$$S_1 = 37 \text{ ppt}; V_1 = 10 \text{ m}^3; S_2 = 12 \text{ ppt}; V_2 = ?$$

$$V_2 = \frac{37 \times 10}{12} = 30.83 \text{ m}^3$$

Freshwater to be added 30.83 - 10 = 20.83 m^3

Example 5: Salinity enhancement

1ppt enhancement = 1g salt content increase is required in one liter of water; and 35 ppt average sea salinity = 35 g salt in one liter of sea water

Preparation of molar and normal solutions

Example 6: (Molar, 1 M: molecular weight in g in one litre of distilled water)

Prepare a solution by dissolving a known mass of solute (often a solid) into a specific amount of a solvent. One of the most common ways to express the concentration of the solution is M or molarity, which is moles of solute per liter of solution.

Prepare 1 liter of 1.00 M NaCl solution.

First calculate the molar mass of NaCl which is the mass of a mole of Na plus the mass of a mole of Cl or $22.99 + 35.45 = 58.44$ g/mol

1. Weigh out 58.44 g NaCl.
2. Place NaCl in a 1 liter volumetric flask.
3. Add a small volume of distilled, deionized water to dissolve the salt.
4. Fill the flask to the 1 litre line mark.

If a different molarity is required, then multiply that number times the molar mass of NaCl. For example, if a 0.5 M solution is required, you would use 0.5 x 58.44 g/mol of NaCl in 1 litre of solution or 29.22 g of NaCl.

Example 7: (Normal, 1 N: equivalent weight in g in one litre of distilled water)

Preparation of N/10 NaOH: Equivalent weight of NaOH is 40, therefore, to prepare N/10, $40/10 = 4$: thus dissolve 4 g of NaOH in one litre of distilled water.

Normality of concentrate acids

Normality of commonly used inorganic acids is as follows -

Concentrate sulphuric acid (H_2SO_4): 36 N

Concentrate nitric acid (HNO_3): 16 N

Concentrate hydrochloric acid (HCl): 12 N

Example 8: (prepare 10% NaCl W/V)

Dissolve 10 g NaCl in 100 ml of distilled water.

Example 9: (prepare 10% formalin V/V)

Mix 10 ml of formalin in 90 ml of distilled water.

Example 10: (N/10 HCl)

Preparation of N/10 HCl: Normality is 12 N, therefore, to prepare N/10, how much distilled water is required?

(i) $N_1V_1 = N_2V_2$ suppose 5 ml of 12 N HCl is taken, then: 12 N x 5 ml = N/10 x V_2

V_2 = 12 N x 5 ml x 10/N = 600 ml; distilled water to be added in 5 ml of 12 HCl = 600 ml – 5 ml = 595 ml.

(ii) First prepare normal (N) solution

$N_1V_1 = N_2V_2$ suppose 1 ml of 12 N HCl is taken, then: 12 N x 1 ml = N x V_2

V_2 = 12 N x 1 ml/N = 12 ml; distilled water to be added in 1 ml of 12 N HCl = 12 ml – 1 ml = 11 ml or 10 ml N HCl + 110 ml distilled water.

Now take 1 ml of this N HCl stock solution

$N_1V_1 = N_2V_2$; N x 1 ml = N/10 x V_2

V_2 = N x 1 ml/N/10 = 10 ml; distilled water to be added in 1 ml of N HCl = 10 ml – 1 ml = 9 ml or 10 ml of N HCl + 90 ml of distilled water.

Example 11: (N/10 HNO$_3$)

Preparation of N/10 HNO$_3$: Normality is 16 N, therefore, to prepare N/10, how much distilled water is required? **First prepare normal (N) solution**

$N_1V_1 = N_2V_2$ suppose 1 ml of 16 N HNO$_3$ is taken, then: 16 N x 1 ml = N x V_2

V_2 = 16 N x 1 ml/N = 16 ml; distilled water to be added in 1 ml of 16 N HNO$_3$ = 16 ml – 1 ml = 15 ml. **Now take 1 ml of this N HNO$_3$ stock solution**

$N_1V_1 = N_2V_2$; N x 1 ml = N/10 x V_2

V_2 = N x 1 ml/N/10 = 10 ml; distilled water to be added in 1 ml of N HNO$_3$ = 10 ml – 1 ml = 9 ml or 10 ml of N HNO$_3$ + 90 ml of distilled water.

Example 12: (N/50 H$_2$SO$_4$)

Preparation of N/50 H$_2$SO$_4$: Normality is 36 N, therefore, to prepare N/50, how much distilled water is required? **First prepare normal (N) solution**

$N_1V_1 = N_2V_2$ suppose 1 ml of 36 N H$_2$SO$_4$ is taken, then: 36 N x 1 ml = N x V_2

V_2 = 36 N x 1 ml/N = 36 ml; distilled water to be added in 1 ml of 36 H$_2$SO$_4$ = 36 ml – 1 ml = 35 ml. **Now take 1 ml of this N H$_2$SO$_4$ stock solution**

$N_1V_1 = N_2V_2$; N x 1 ml = N/50 x V_2

V_2 = N x 1 ml/N/50 = 50 ml; distilled water to be added in 1 ml of N H_2SO_4 = 50 ml – 1 ml = 49 ml or 10 ml of N H_2SO_4 + 490 ml of distilled water.

Note: Always add acid into water drop by drop; never add water into acid as it causes the strong exothermic reaction.

Example 13: (1.5 ppm iron removal from 10 M^3 available source water; iron content x 0.64 chlorine)

Total amount of water for chlorine removal = 10 M^3 x 1000 L = 10,000 L

Total iron concentration = 1.5 ppm x 10,000 L = 15,000 ppm

Total chlorine requirement = 15,000 x 0.64 ppm = 9600 mg = 9.6 g

Approximate chlorine availability in bleaching powder = 30%

Bleaching powder requirement = 9.6 x 100/30 = 9.6 x 10/3 = 32 g.

Example 14: (1.5 ppm formalin application in 10 M^3 freshwater prawn hatchery tank infested with ciliate protozoan, *Zoothamnium*)

Total amount of water for treatment = 10 M^3 = 10 x 1000 L = 10,000 L

Formalin requirement @ 1.5 ppm = 1.5 ml/M^3

Total formalin requirement @ 1.5 ppm = 1.5 ml x 10 M^3= 15 ml.

Example 15: (0.05 ppm copper sulphate application in 50 litre fungal infested aquarium)

Prepare stock solution: dissolve 1 g copper sulphate in 1 litre of distilled water = 1 g/l = 1 mg/ml

Total copper sulphate requirement; 0.05 ppm x 50 L = 2.5mg/50L

Administer in aquarium 2.5 ml of stock solution prepared.

Example 16: (0.01 ppm x alkalinity concentration; copper sulphate application in 0.5 ha area, average depth of 1.0 M, to control algal eutrophication)

Suppose alkalinity concentration is 325 ppm

Total water volume = 0.5 ha x 1.0 M = 5000 M^2 x 1.0 M = 5000 M^3

Copper sulphate requirement; 0.01 x 325 mg = 3.25 mg/L = 3250 mg/M^3 = 3.25 g/M^3

Total copper sulphate requirement = 3.25 g x 5000 M^3 = 16250 g = 16.25 Kg.

Example 17: (Lime application @ 3 ppm in 1ha area, average depth of 1.5 M)

Total water volume = 1 ha x 1.5 M = 10,000 M^2 x 1.5 M = 15,000 M^3

Lime @ 3 ppm in one litre is 3 mg or in 1 M^3 = 3 mg x 1000 L = 3000 mg = 3g

Total lime requirement = 3 g x 15,000 M^3 = 45,000 g = 45 Kg.

Example 18: (Mn, manganese @ 0.1 ppm to remove ciliate infection in 1ha area, average depth of 1.0 M)

Total water volume = 1 ha x 1.0 M = 10,000 M^2 x 1.0 M = 10,000 M^3

Mn @ 0.1ppm through $KMnO_4$ application

0.1 ppm = 0.1mg/l or 100 ug/l

Mn for 1000 litre or 1 M^3 = 1000 x 0.1 mg = 100 mg

Mn for 15000 M^3 = 100 mg x 15000 = 1500000 mg

Total Mn requirement = 1500000 mg /1000 = 1500 g = 1.5 Kg

Total $KMnO_4$ requirement = 158x1.5/55 = 4.3 Kg, where 158 is molecular weight of $KMnO_4$ and 55 is atomic weight of Mn.

Example 19: (Sodium nitrate, $NaNO_3$ application to control burgeoning TSS quantum/viscosity and reciprocally depleting DO in hatchery waters)

Total amount of water for treatment = 10 M^3 = 10 x 1000 L = 10,000 L

Oxygen requirement @ 5 ppm = 5 mg/litre = 5000 mg/M^3 = 5 g/M^3

Total dissolved oxygen requirement @ 5 ppm = 5 mg x 10 M^3 = 5 mg x 10 x 1000 L = 5 g/M^3

$\qquad$ = 50 g/10 M^3

Molecular weight of $NaNO_3$ = 23 + 14 + 48 = 85

48 g oxygen needs 85 g of $NaNO_3$, therefore 50 g oxygen needs how much $NaNO_3$?

$NaNO_3$ = 50 x 85/48 = 88.54 g

Example 20: (Hydrogen peroxide, H_2O_2 application to enhance DO in intensive systems)

Total amount of water for treatment = 10 M^3 = 10 x 1000 L = 10,000 L

Oxygen requirement @ 5 ppm = 5 mg/litre = 5000 mg/M^3 = 5 g/M^3

Total dissolved oxygen requirement @ 5 ppm = 5 mg x 10 M^3 = 5 mg x 10 x 1000 L = 5 g/M^3

= 50 g/10 M^3

Molecular weight of H_2O_2 = 2 + 32 = 34

32 g oxygen needs 34 g of H_2O_2, therefore 50 g oxygen needs how much H_2O_2 ?

H_2O_2 = 50 x 34/32 = 53.125 g

But now generally, the commercially available different hydrogen peroxide brands are with 50% available oxygen (w/w), therefore, to make available 50 g oxygen; 100 g hydrogen peroxide is to be applied.

Example 21: (0.05 ppm iodine application in 1M³ water as antimicrobial dose)

Total iodine requirement: 0.05 ppm x 1000 litre = 50 ug x 1000 = 50,000 ug = 50 mg

Prepare stock solution: dissolve 1 g iodine in 1 litre of distilled water = 1 g/l = 1 mg/ml

Administer 50 ml of stock solution prepared.

Example 22: (Magnesium supplementation in 10 M³ inland saline water through anhydrate $MgSO_4$)

Total amount of water for treatment = 10 M³ = 10 x 1000 L = 10,000 L

Suppose salinity concentration of inland saline water is 12 ppt and available magnesium is 240 ppm.

Required magnesium at the benchmark of 35 ppt @ 1350 ppm in 10 M³ water of 12 ppt salinity

$$= \frac{12 \text{ ppt} \times 1350 \text{ ppm}}{35 \text{ ppt}} = 462.857 \text{ ppm}$$

Required magnesium in one litre = 462.857 mg

Therefore in 10 M³ or 10,000 L = 462.857 x 10,000 = 4628570 mg = = 4628.570 g = 4.6285 Kg

Available Mg in 10 M³ inland saline water @ 240 ppm = 240 mg x 10,000 L = 2400000 mg = 2400 g = 2.4 Kg

Amount of magnesium supplementation = 4.6285 Kg - 2.4 Kg = 2.2285 Kg

Molecular weight of anhydrate $MgSO_4$ = 24.32 + 32.06 + 16 x 4 = 120.38

Total anhydrate $MgSO_4$ requirement $= \dfrac{2.2285 \times 120.38}{24.32} = 11.03 \text{ kg}$

Note

ppm for –

1. Solid – solid phase : mg/Kg
2. Solid – liquid phase : mg/l
3. Gas – gas phase : μl/l

CHAPTER 17

ENVIRONMENTAL STRESS AND REDOX POTENTIAL

INTRODUCTION

Fishes are aquatic and poikilothermic animals. Their existence and performance are dominated by the quality of their environment. While conditions in large water bodies like reservoirs, lakes, seas are relatively uniform, smaller bodies of water such as ponds, tanks, hatcheries *etc.* in which fishes are raised, are subject to more variable conditions. The fish farmers carry out many procedures like liming, manuring, netting, grading, transporting, stocking, feeding, sampling, treating, transfer from freshwater to brackish water or vice versa which may cause stress to fish in some way or other. As the intensity of culture becomes higher to achieve more fish production, the stocking density, fertilization, feeding rate and other associated husbandry procedures also increase and thus resulting in ever increasing stress on fish in turn deteriorating the water quality leading to the outbreak of diseases.

Stress

In medical terminology, stress is a body condition that occurs in response to actual or anticipated difficulties in life. Stress was defined by Selye (1936) as the, "sum of all the physiological responses by which an animal tries to maintain or reestablish a normal metabolism in the face of a physical or chemical force". This definition, however, does not consider the fact that the outcome of stress may be negative for an individual but positive for the population when space and food supplies are limiting. Further, stress in fish can be characterized by physiological changes such as plasma cortisol, glucose, lactate and electrolyte concentrations and is quantitatively related to the severity and longevity of the stressor. Stress is better defined as the, "effect of any environmental alteration or force that extends homeostatic or stabilizing processes beyond their normal limits, at any level of biological organization". In another way, stress can be defined as, "any biotic and abiotic factor that produces significant disturbances in the normal functions of an animal and thus decrease the probability of survival". The definition of Brett (1958) for fishes is similar and provides a quantitative measurement of applied stressor, "stress is a state produced by an environment or other factor which extends the adaptive responses of an animal beyond the normal range or which disturbs the normal functioning to such an extent that, in either case, the chances of survival are significantly reduced".

Environmental stress and fish diseases

The fact that various kinds of environmental changes are stressful and lower the resistance of fishes to infections and other diseases have become better understood in recent years. Experience from fish farming and hatchery management have shown that a wide variety of bacterial, parasitic or other diseases become a problem only if fish are held under environmental conditions unfavourable for that particular species. Such unfavourable conditions include overcrowding, temperature fluctuations, salinity fluctuations, inadequate dissolved oxygen, high concentrations of obnoxious gases like hydrogen sulfide, carbondioxide and ammonia, excessive or rough handling, sublethal levels of toxic materials *etc.* Examples of stress mediated diseases in warm and cold water fish culture include vibriosis (*Vibrio anguillarum*), gill disease (*Myxobacteria* sp.), columnaris disease (Kumar *et. al.*, 1996), bacterial haemorrhagic septicemia (*Aeromonas hydrophila)* and such protozoan diseases as Ichthyobodiasis (*Ichthyobodo necatrix*) *etc.,* (Wedemeyer, 1970).

Fish in high intensity aquaculture are continuously affected by environmental fluctuations. Unscientific management practices, together with resulting fright, can impose considerable stress on the limited homeostatic mechanism of most fishes. Stress requiring an adjustment that exceeds a fish ability to accommodate becomes lethal. Less severe stress may predispose the fish to physiological or to infectious diseases if pathogens are present (Wedemeyer, 1970). Coagulated yolk, blue sac and hauling loss are examples of physiological problems caused by excessive environmental stress. Bacterial gill disease (*Myxobacteria*), haermorrhagic septicemia (*Aeromonas* and *Pseudomonas spp.*) and furunculosis (*Aeromonas salmonicida*) are infectious diseases of cultured fish known to be caused by adverse environmental changes.

Physiological response of fishes to stress

Stress modulates many of the defense mechanisms, suppressing some and exaggerating others. This may be advantageous or injurious to animal concerned depending upon the complex interactions between the stress factors and the animal's physiological state, which eventually determines how well the animal will adapt to the situation. The classical work on the morphological and physiological responses of animals to stress is undoubtedly that pioneered by Selye (1936, 1946, 1950, 1956). According to him, a series of morphological, biochemical and physiological changes occur as a result of stress which is collectively termed as General Adaptation Syndrome (GAS). Selye (1973) divided GAS into three stages: (a) Alarm reaction, (b) The stage of resistance during which adaptation to stress has occurred to achieve homeostasis under the changed circumstances and (c) The stage of exhaustion when adaptation has been lost because the stress was too severe or long lasting and the fish fails to achieve homeostasis. These stages are characterized by a variety of metabolic changes which are not species specific but are remarkably similar for any stressing agent, for example, anoxia, infection, fright, forced exertion, anaesthesia, temperature changes and injury *etc.*

Host-parasite-environment relationship

'Parasite' is an organism that feeds and lives on any other living organism, the latter being called 'host'. The parasites feed on host's tissue(s) to sustain themselves and their wastes are discharged directly into the host body. 'Environment' is everything that is external to an organism. They are either abiotic or biotic. The abiotic environment is made up of many factors and forces that influence one another and the surrounding community of living things. They include temperature, sunlight, soil, water, atmosphere and chemical factors. The biotic or living environment includes fish food, plants and animals and their interactions among one another. An organism's survival and well-being depends upon food and association with other living things. The actual initiation of disease in fishes or for that matter in any higher vertebrate is a complex process which involves more than mere contact between the host, parasite and environment. Host susceptibility, parasite's virulence and environmental factors must all interact. Thus saprophytic bacteria such as *Pseudomonas fluorescens*, *Aeromonas hydrophila* (bacterial septicemia), *Myxobacteria sp.* (bacterial gill disease), *Vibrio anguillarum* (vibriosis in marine fishes) are normally present in natural waters at all times but do not usually cause disease problems unless environmental stress of some type intervenes.

Stress factors causing outbreak of diseases in fish

Higher fish production can be achieved if the physical, chemical and biological properties of the water in the ponds are maintained at optimal level. There are many stress factors which in either excessive or deficient quantity can adversely affect the fish and predispose them to disease outbreak. Broadly, they are of four types: (A) Physical, (B) Chemical, (C) Biological and (D) Procedural.

PHYSICAL STRESS FACTORS

Water temperature

The health of individual fish or of a given species in high density aquaculture is greatly influenced by its ability to adapt to variations and extremes in ambient water temperature. Each species is characterized by a range of temperature in which it thrives. Sustained temperature above or below this range, or rapid changes within this range represent stressful environmental conditions.

The gradual changes in water temperature seldom cause problems in fish health; however, sudden change may impose severe stress. Such temperature induced stress may be sufficiently more to lower the disease resistance of fish and increase their susceptibility to infections. Thread fin shad experience mass mortalities when sudden temperature drops occur even when well above the lower lethal temperature. Subclinically infected shad are much more sensitive to temperature changes than healthy fish and mortalities due to activated *Aeromonas hydrophila* infections as reported by Haley *et al.*, 1967. Alanson *et al.* (1971) demonstrated that *Tilapia mossambica* got disoriented when the temperature was slowly lowered from 25°C to 11°C and became comatose after 5 days at that temperature. They concluded that death occurred due to osmoregulatory collapse and renal failure.

Increase in temperature adversely affects the health of fish and other aquatic life by lowering the dissolved oxygen content of the water, increasing oxygen demand by enhancing metabolic respiratory rates, increasing the toxicity of harmful substances dissolved in water affecting growth, invasiveness and virulence of bacterial and other pathogens and thus lowering the disease resistance of fish. In furunculosis disease of salmon, it was found that higher water temperature not only increased total mortality but also shortened the time period between infection and death (Wedemeyer *et al.*, 1999). The *Flexibacter columnaris* infection in salmonids shows similar pattern. In temperate aquaculture, water temperature below 10°C seems to enhance fish survival while above this mortality becomes progressively greater with each incremental temperature increase. Surviving fish do not completely eliminate *F. colmnaris* from their tissues and if the water temperature then gets increased to about 18°C, fatal infection can develop. Fryer and Pilcher (1974) also showed that infections of salmon with the protozoan parasite, *Ceratomyxa shasta* coupled with increase in the water temperature shortened the time from infection to death. Further, the combination of stress-related elevation in energy demand and reduction in feeding due to cold temperature and short photoperiod, leading to severe depletion of stored body lipid; coined with the name Winter Stress Syndrome. Thus, temperature is the etiological agent of disease not only when conditions deteriorate beyond the limits tolerated by the species involved but also, it is an important contributing factor when water temperature causes stress in the presence of fish pathogens or when the fish are carriers or with latent or subclinical infection.

CHEMICAL STRESS FACTORS

1. pH

Low pH increases the susceptibility of fishes to diseases. Neess (1949) reported that below pH 5.5, carp develop hypersensitivity to aquatic bacteria. Acidic pH alters the ionic concentration in fish and buffering capacity of water. In acidic water, *Catostomus commersoni* showed reduced feeding response and weight loss (Beamish, 1972). Low pH decreases fecundity and egg fertility in fishes (Menendez, 1976). Schaperclaus (1991) reported that if the pH value of pond water decreases to 4.8-5.0 for some time, carp succumb to microbial or parasitic infections. In acidic water fish secrete more slime and thus in turn become more hyper sensitive to microbial infections and sometimes show the decolouration of the gill tissue.

Viability of fish spawn is largely dependent on pH. Experiments with the eggs of perch and roach have shown that entire roach spawn perished at pH 4.7, while perch at the same pH, had a hatching success of only 28%. At pH 5.7, 50% of the perch survived the early fry stage. On the other hand, only 14% of the roach got alive (Milbrink and Johansson, 1975). The increase in pH above the optimum range can cause as much stress to fish as the acidic waters. Highly alkaline waters corrode the gills and fins rays. Eggs of trout perish when pH values go above 10. The eggs swell and yolk takes on a whitish colour. At pH 11, cornea of many fish becomes opaque, many of the survivors become blind in one or both the eyes (Seymour and Donaldson, 1953). In addition to this, a change in water pH either to acidic or alkaline conditions exerts stress in fish which can be characterized by swelling of erythrocytes, production of immature erythrocytes, reduction in the total erythrocyte counts, haemoglobin and serum protein contents.

2. Oxygen deficiency

Fish culture ponds are often imperiled by the problems of oxygen deficiency because the capacity of water to hold dissolved oxygen is small. Whereas one liter of air contains 210 cm^3 of O_2 and 790 cm^3 of N_2 (including small quantity of rare gases), the oxygen content of water amounts to a small percentage of atmospheric content and one litre of water at 20°C contains only 6.3 cm^3 of oxygen (= 8.9 mg of O_2). The absorption coefficient for oxygen at 20°C is only about 1/32 and it is temperature dependent. The absorption coefficient decreases with increasing temperature, so the oxygen value under these conditions decreases and vice versa. If the oxygen is used up it can only be absorbed relatively slowly from the atmosphere by the calm water surface and diffused into the deeper water layers. Maximum oxygen solubility in water at -4^{0}C is 16 ppm.

Oxygen requirement of different species of fish is fairly variable and depends on several factors like fish weight, water temperature, intake of feed and their activities. When there is deficiency of oxygen, the defence mechanism of fish is no longer maintained at optimum level. The possibility of parasitic infection particularly of endoparasites increases, since they mostly reside in the body cavity and tissues; and are, therefore, adapted to an environment of poor oxygen. Under oxygen deficiency, the number of erythrocytes rises at first and drops thereafter. The content of plasma proteins also decreases and along with it the ability to form antibodies. Lloyd (1961) reported that low environmental oxygen concentration may enhance the stressful effects of other toxicants or increase the probability of disease.

Egg maturation and later egg development are particularly endangered by oxygen deficiency. Inadequate supply of oxygen hampers ovulation and in later stage results in death of embryos. If oxygen concentration in the water is low even after hatching, the fry have poor growth. Einsele (1956) reported that fry of *Salmo gairdneri* are affected at 5 mg/l of oxygen, a risk of asphyxiation arises at 3 mg/l and mortality occurs at 2 mg/l. Though many fishes can physically survive low oxygen levels, at least for limited periods, almost any reduction below oxygen saturation will cause some adverse effects on growth, reproduction, activity or other physiological functions. Although there is probably no single upper limit on dissolved oxygen concentration which applies to all fishes or fish cultural situations, there is general agreement on the lower limit: 4 ppm for warm water species and 5 ppm for salmonids below which conditions are decidedly unfavourble for their sound health.

3. Carbon dioxide complexities

Carbon dioxide is a natural constituent of air and it is 28 times more soluble in water than oxygen. It occurs in water mainly as a product of organic decomposition. Carbon dioxide can be found in three closely related forms in water: CO_2 (dissolved free carbon dioxide), HCO_3^- (bicarbonate ion), and CO_3^{--} (carbonate ion), the concentration of each one depends on pH level. In aquahatchery and aquaculture practices, CO_2 concentration should not exceed more than 3 ppm. In culture ponds, for the process of photosynthesis and natural fish food production; carbon concentration is required, which is supplemented by CO_2, HCO_3^- and CO_3^{--}. Therefore, CO_2 level at 3 ppm and alkalinity range between 50-250 ppm is desirable to have better aquatic primary productivity.

Table 34: Correlation of CO_2 solubility in relation to pH

Parameter	% of total CO_2 in each form in relation to pH							
pH	4	5	6	7	8	9	10	11
CO_2	99.5	95.4	67.7	17.3	2.0	0.2	–	–
HCO_3^-	0.5	4.6	32.3	82.7	97.4	94.1	62.5	14.3
CO_3^-	–	–	–	–	0.6	5.7	37.5	85.7

When CO_2 gets dissolved in water, H_2CO_3 is formed which then dissociates into HCO_3 if the water is alkaline. The opposite occurs and most of the bicarbonate is converted to H_2CO_3 in acidic water. H_2CO_3 is thermodynamically unstable and decomposes into CO_2 and H_2O. Carbon dioxide can act as a limiting factor for example, during transport of fish or in maintaining fish in closed systems with recycled water and in borewell-water-fed aquaculture where CO_2 concentration may sometimes be very high. In the active fish, as the CO_2 concentration increases, the metabolic rate declines, $i.e.$, the oxygen consumption goes down. Since CO_2 decreases the affinity of blood for oxygen and reduces its carrying capacity, an increased CO_2 content of water raises the asphyxation threshold of fish.

The desirable amount of free CO_2 for optimum fish rearing and culture practices ranges between 1.5 to 3.0 mg/l. Detrimental effects of CO_2 on freshwater fish begin above concentrations of about 3.0 ppm and anaesthesia occurs at around 10 ppm. Schaperclaus (1991) reported that rainbow trout which are bred in hatcheries with high CO_2 content in water many times suffer from nephrocalcinosis. On the other hand, too low CO_2 content in water can also be fatal for carp fry. If the amount of free CO_2 in the water is lesser than 1 mg/l, this leads to hyperventilation syndrome in carp larvae, which respire through the entire body surface. The consequence is a strong demand on the buffering capacity of blood which results in a destabilization of the acid-base balance and during their maintenance there is a respiratory alkosis, or after the first uptake of protein-rich food there is excessive metabolic acidosis (Schaperclaus, 1991). However, dissolved CO_2 can also be useful in pond culture as it decreases the concentration of unionized ammonia because of drop in pH.

4. Nitrite toxicity

Nitrites are ubiquitously distributed in natural water bodies. Nitrites arise either by denitrification of nitrates or by nitrification of ammonium ions.

$$NO_3^- \rightarrow NO_2^- + \frac{1}{2} O_2$$

$$NH_4^+ + 2O_2 \rightarrow NO_2^- + 2H_2O$$

Wedemeyer and Yasutake (1978) showed that toxicity of nitrite is determined by nitrous acid (HNO_2) as a consequence of dissociation constant of nitrite depending on the pH, temperature and water hardness.

$$NO_2^- + H_2O \rightarrow HNO_2 + OH^-$$

In fish, HNO_2 toxicity leads to methemoglobinemia, which with a methemoglobin content of 5% of the haemoglobin content leads to physiological oxygen deficiency. Damage resulting from HNO_2 can produce a violet colour in gills, hypertrophy of gill epithelium, metabolic alkalosis and dyspnoea.

5. Ammonia toxicity

Ammonia occurs naturally in ground and surface waters, primarily as a result of microbial decomposition of nitrogenous organic substances in the ponds like metabolic products of organisms, oxidation-reduction of fertilizers and wastewaters *etc*. Effluents may also add ammonia. Besides, excretion by aquatic organisms, particularly fish and shellfish in intensive aquaculture systems also generate ammonia. Death of phytoplankton blooms also gives rise to high levels of ammonia in aquatic systems. It is the byproduct in nitrogen cycle through the processes of denitrification, wherein ammonia is oxidized to nitrite and nitrate in oxygenated waters; and in deoxygenated waters, nitrate is converted to nitrite and ammonia (denitrification). Thus, deoxygenation in heavily loaded ponds such as intensive aquaculture systems, can therefore lead to a buildup of ammonia. It is also generated during fish and fish seed transportation. The ammonia toxicity is the highest of all the compounds of nitrogen. The poisonous effect of ammonia in water depends decisively on the pH value. Higher the pH of water, higher is the ammonia toxicity and vice versa. The percentage of unionized ammonia in water of different pH and temperature is presented in table below.

Table 35: Correlation of ammonia solubility in relation to pH and temperature

pH	Temperature (°C) unionized ammonia (%)	
	20	32
7.0	0.4	1.0
8.0	3.8	8.8
8.2	5.9	13.2
8.4	9.1	19.5
8.6	13.7	27.7
8.8	20.1	37.8
9.0	28.5	49.0
9.2	38.7	60.4
9.4	50.0	70.7
9.6	61.3	79.3
9.8	71.5	85.8
10.0	79.9	90.6
10.2	86.3	93.8

Further, it is to be noted that the unionized ammonia (NH_3 - N) is toxic to fish at the level of 0.1 ppm while the ammonium nitrogen (NH_4^+ - N) at 1.0 ppm. The percentage of unionized ammonia in a given sample of total ammonia can be calculated from the equation:

$$pKa = 0.09018 + \frac{2729.92}{(273 + T)}$$

Where

T = temp (°C)
UIA = unionized ammonia
pKa = negative base -10 logarithm of the acid dissociation constant (Ka) of a solution

$$\% \ UIA = \frac{100}{(1 + anti \log (pKa - pH))}$$

The damage due to ammonia is confined to respiratory organs, blood and nervous tissues of fish. The NH_3 molecule has high permeability. Therefore, during ammonia intoxication from water, ammonia can penetrate the blood and tissues via gills and skin or can get accumulated in the organism to a toxic concentration. Schaperclaus (1991) reported that under the action of acute toxic ammonia on fish, there is considerable slime secretion, swelling of the epidermis and hyperemia in the entire body surface or only on gills. Frayed and pellucid fin ends and epithelial detachment are noticeable. The gills show hyperplasia and in the advanced stage, haemorrhage, histolysis and necrosis occur. The liver, spleen and kidney become pale and there are swelling of the parenchymatous cells, haemorrhage, inflammations and other degenerative processes. A fall in the number of erythrocytes in the haemoglobin content can be observed in the blood, but no significant change takes place in the number of leucocytes. When there are acute NH_3 intoxications, there appear shock symptoms, muscular twitches, and disorders of the ocular rotatory reflexes, rotation and non-directional swimming. The chronic effect of sublethal NH_3 concentrations results mainly in high susceptibility to ectoparasitosis and to myxobacterial infections. The oxygen content in the tissue drops with increased NH_3 concentration in fish.

6. Hydrogen sulfide toxicity

Hydrogen sulfide (H_2S) is colourless and highly toxic gas which smells like rotten eggs. Hydrogen sulfide in pond water originates from reductive biochemical transformation of inorganic and organic substances which contain sulfur. Besides, it also comes from the effluents of sewage treatment plants, chemical and metallurgical industries *etc*. Hydrogen sulfide exists in two forms in water: HS (ionized sulfide ion) and H_2S (unionized hydrogen sulfide gas). More than 0.003 ppm of unionized hydrogen sulfide gas is highly toxic to fish. Like ammonia-ammonium the percentage fractions of hydrogen sulfide depends directly on pH of the water. But in this case, lower the pH, higher is the hydrogen sulfide and vice versa. Raising pH value reduces the toxic effect of H_2S.

$$H_2S \rightleftharpoons HS^- + H^+$$

Hydrogen sulfide is highly toxic to fish whereby respiratory frequency and intake of oxygen are considerably reduced during exposure. Under action of hydrogen sulfide, an increasing proportion of malformations occur in the hatchlings. The larvae in the yolk sac stage, when exposed to higher concentration of H_2S, become smaller than those of the control. Chronic exposure of high sublethal concentrations of H_2S, retards the growth in blue-gill. Kumar *et al.* (1996) reported hydrogen sulfide mediated columnaris disease in Indian Major Carps. The toxicity of H_2S rises as the oxygen content decreases. The average value of LC_{50} decreases from 71 to 53 µg/l (Adelman and Smith, 1972). An increase in water temperature and decrease in oxygen concentration enhance the toxicity of H_2S. EPA (1973) envisages that H_2S concentration in fish ponds should not exceed 0.002 ppm.

7. Iron toxicity

Although little iron is found in solution in surface waters as it is usually oxidized to insoluble ferric hydroxide and gets precipitated, however, ground waters may contain substantial amounts of iron in ferrous forms, usually as ferrous bicarbonate. Water can dissolve greater amounts of ferrous bicarbonate when it is nearly free of dissolved oxygen; it contains substantial amount of carbon dioxide, when, the pH is not above 7.5, and the organic decomposition products are present which can reduce ferric hydroxide. While the direct toxicity of iron and its salts is generally low, lethal effects have been noted following exposure of fish to ferrous iron in poorly buffered, low pH water (McKee and Wolf, 1963). Indirect lethal effects have been attributed to the precipitation of ferric hydroxide or ferric oxide on the gills of fish (Lewis, 1960). Precipitation of ferric hydroxide on incubating eggs may also smother and kill developing embryos. Schaperclaus (1991) reported that iron precipitates cause necrosis of the gills of young trout.

8. Heavy metal toxicity

While metals themselves are hardly toxic for aquatic organisms, their salts in solution are highly toxic to aquatic life. Heavy metals may alter the physiological activities and biochemical parameters both in tissues and in blood of aquatic organisms. Further, heavy metals often induce a delay in the hatching process, premature hatching, deformations and death of newly hatched larvae. Dissolved metallic compounds occur in the effluents, sewage systems, paper factories, electroplating and metal processing industries. The visible effects of toxicity of metal ions in fish are characterized by restlessness, increased or marked decreased frequency of opercular beatings, secretion of slime, reduced sensitivity to stimulation, exhaustion, lateral disposition and death due to asphyxiation. The toxicity of salts of heavy metals to fish is intensified by a drop in the dissolved oxygen content of water. Increasing water hardness, in which many metal ions are precipitated, results in a reduction of the toxicity, while in soft water (100 ppm as $CaCO_3$, or less), they are comparatively more toxic.

BIOLOGICAL STRESS FACTORS

1. Algal toxicosis

With increasing nutrient enrichment of water bodies, there is a corresponding increase in the incidence of large scale algal blooms. The incidences of oxygen depletion caused mass mortalities of fishes are well known. On the other hand, toxic algal blooms are less conspicuous and little difficult to recognize. Certain species of freshwater algae, if present in large quantities produce powerful toxins which can seriously affect aquatic life. Blooms of these algae frequently occur in highly fertile ponds or in areas of organic pollution, but they may also develop under normal feeding programmes or following application of fertilizers. But, the recognition of toxic conditions often goes unrecorded since such heavy algal blooms often terminate in an oxygen depletion which is then turned to cause fish mortality. Normally, no efforts are taken to detect the toxic algae and their toxic effects. However, the differences between the algal toxicosis and oxygen depletion are quite significant and the recognition of this may provide information for corrective action.

Toxic effects on fish are known from following species of algae: *Anabaena flosaquae, Microsystis aeruginosa, Aphanizomenon flosaquae, Nodularia sp., Chlorella vulgaris, Prymnesium parvum, Glenodinium foliaceum, Ochromonas sp., Peridinium polonicum, Gymnodinium brevis, G. flavum, Gonyaulax polyedra, G. catenella, Exuviella baltica* and *Coelosphaerium sp.* Fishes subjected to algal toxicosis exhibit erratic swimming behaviour, become lethargic, then convulsive and ultimately may die. Smaller and less hardy fish species usually die first but entire pond fish populations may ultimately be killed. For example, the effects of cyanobacterial blooms containing *Microcystis sp.* on fresh water fish, *Heteropneustus fossilis* were investigated under laboratory conditions which resulted an increase in bilirubin concentration in the blood of treated fish. Further, haemoglobin level and cholesterol concentration get also altered due to toxicity of *Microcystis sp.* However, a Euglenoid, *Euglena sanguinea* is reported to produce a potent *Ichyotoxin* (an alkaloid) which is similar in structure to fire ant venom.

2. Overcrowding

Overcrowding means stocking of a large number fish in a limited space of per unit area exceeding the carrying capacity of a particular water body. Crowding has been recognized as one of the most important stress factors predisposing fish to various disease outbreaks. The number of fish which can successfully live and grow in a given water body depends on the area of the pond, dissolved oxygen level, metabolic rate of fish, available fish food, pathogenic load and the frequency of water exchange. Crowding creates lots of social stress among competing fish species for food, space and dissolved oxygen. The effects of high stocking density include decreased growth, reductions in food intake, food conversion efficiency, nutritional status, fin erosion, gill damage, reduced immune capacity and alterations in swimming behaviour.

The carrying capacity is based on oxygen requirement which is directly proportional to water temperature and the size of fish. Crowding greatly facilitates the horizontal transmission of contagious diseases of fish. High population density can also lead to diseases of indirect cases. The high density cage-culture of fish may result in over browsing of natural source of vitamins and essential nutrients within the cages. The broken-back syndrome which occurs in cage-reared channel catfish reflects a deficiency in the availability of natural source of vitamin C (Lovell, 1973). Unless supplemental vitamin C is added to the diet, caged fishes may develop fractures in the vertebral column. Fish which are permitted to roam free in the waters, find ample vitamin sources to meet their needs. Reducing population density improves environmental conditions for the fish and makes parasite transmission among fish less probable.

Procedural stress factors

Procedural stress factors include stocking, handling, hauling, treatment procedures and feeding methods. They are basically anthropogenic in nature. These stress factors are mostly ignored during culture operations but they can cause serious outbreak of diseases by activating latent infections.

Redox potential (Eh)

The redox potential is a measure of reducing or oxidizing strength of a solution. It is an important parameter affecting metal and nutrient solubility and mobility. A negative potential value means that the solution has a reducing action when compared with a standard hydrogen electrode. A positive value means that the solution has an oxidizing effect. The redox potential system in natural waters is considerably influenced by temp. and pH. The potential measured at the prevailing pH is adjusted to pH 7. This measurement is called Eh7. When so adjusted, the observed potentials generally lie between 0.4 volt (400 mv) and 0.5 volt (500 mv), which are thus slightly lower than the potential of 0.52 volt at 25°C; the best value for aerated water at one atmospheric pressure. Anaerobic bacteria exist where Eh lies below 0.4 volt.

Healthy aquaculture systems should have a redox potential ranging between 350 to 400 millivolt. Levels below 200–250 mv indicate the presence of toxic reduced compounds. While levels above 520 mv indicate too active oxidative environment, which can damage plant and animal cells and tissues. Overtreatment with ozone can produce too high redox potential. It can be maintained at a comfortable 350-400 mv with careful ozone regulation. General effectiveness ratio of oxygen: chlorine: ozone is 1: 2: 3.

*The 22.4 litres of any gas will weigh (in g) the same amount as its molecular weight at normal atmospheric pressure and 0^0C temperature. Therefore, 22.4 litres of oxygen, chlorine and ozone will weigh 32 g, 71 g and 48 g respectively.

Table 36: Redox potential and prevalent oxygen (mg/l)

Sl. No.	Reaction	Redox potential (mv)	O_2 (mg/l)
	Reduction		
1.	NO_3 to NO_2	450 - 400	4.0
2.	NO_2 to NH_3	400 - 350	0.4
3.	Fe^{+++} to Fe^{++}	350 - 200	0.1
4.	SO_4^- to S^-	100 - 60	0.0
	Oxidation		
5.	$O_2 + H_2O = HO + HO_2$	400 – 500	5.0
	$O_3 + H_2O = HO_3 + OH^-$	> 500	> 5.0

CHAPTER 18

DESIGNING OF EFFLUENT TREATMENT PLANTS

INTRODUCTION

The Aquaculture Authority has made it mandatory that all shrimp farms of 5.0 hectare water spread area and above located within the CRZ and 10 hectares water spread area and above located outside CRZ should have an effluent treatment system (ETS) or effluent treatment ponds/ facility. Establishment of such a system is necessary to bring the shrimp farm wastewater within the prescribed standards and mitigate any adverse impact on the ecology of the open waters. An effluent treatment system consisting of settlement/sedimentation pond (SP), bio-ponds (BP) and aeration pond (AP) is proposed for the shrimp farms practicing improved traditional and intensive methods of farming. By incorporation of the ETS facility, the farm wastewater is expected to be as good in quality as that of intake water. Quality-wise, the treated wastewater would also be suitable and ideal for recirculation within the farm, making the farming practice conform to the zero discharge norms. However, such a system (*i.e.,* recirculation system) would need the establishment of a reservoir pond of suitable size. Thus, effluent treatment plant is a process design to treat wastewater for its safe disposal and/ or its reuse.

Effluent treatment system– design and layout

Science is universal and technology is regional. Ideas make the plan and to sketch the plan is called as design, and its implementation is to be done through engineering and technological inputs. Pollution control is an essential task. There are four types of control; legal, social, economical and technological. Legal, social and economical measures help to prevent the pollution from occurring; while technological measures make it easier to control the existing pollution.

Basic considerations

The characteristic features of the shrimp farm wastewater that have been taken into consideration for designing the ETS are as follows:

- Shrimp aquaculture wastewater/discharge during normal operating conditions (*i.e.*, during culture period) is high in volume, but relatively dilute in nature. When ponds are aerated (farms adopting improved traditional and extensive methods of farming), discharge from the ponds usually contains adequate oxygen for aquatic life and diluted concentrations of nitrogen, phosphorous and organic matter.

- The shrimp pond water quality tends to deteriorate through the grow-out period, as feeding rate increases with shrimp size and biomass. Thus, the highest quantity and poorest quality of wastewater (in terms of nutrients load, total ammonia, ionized ammonia and total suspended solids) is found just before harvest time, when shrimp biomass is at the maximum.

- Wastewater discharge during harvest (especially the last 5 cm drainage) is usually the most important contributor to overall wastewater loading, comprising over 75% of the total load.

- Stocking densities and management practices largely influence the quality of the discharge. In areas where the intake water quality is below the desired standards, a recirculation system can be resorted to use a storage reservoir in combination with the ETS. This would help in maintaining the desirable water quality and animal health standards.

Lay-out plan

The lay-out plan of the ETS is depicted in the figure 24. As per norms, 10 per cent of the cultivable area should be assigned for the ETS. For example, for a farm of 5.0 hectare water spread area, 0.77 hectare land area or approximately 0.50 hectare water spread area (actual operational area) will be required for construction of the ETS. For farms more than 5.0 hectares, the area under ETS will also proportionately increase (*e.g.*, for a 6.0 ha. farm area, 0.6 hectare under ETS; for a 10 hectare farm area, 1.0 hectare under ETS and so on). The size of the sedimentation/settlement pond, biopond and aeration pond has also been suggested taking into consideration the optimum production level of 2.0 tonnes/hectare/culture and specific water management practices. Water exchange schedule to be followed for operating the system is shown below. The schedule is based on the availability of a reservoir of suitable size for storage and treatment of water for initial filling of the ponds, topping up of water level during the first two months of rearing and limited water exchange during the third and fourth months of rearing. Even though, a modular design of ETS is shown, the design and lay-out may be suitably altered taking into consideration the location, contour interval index and topography of the land available for such purpose.

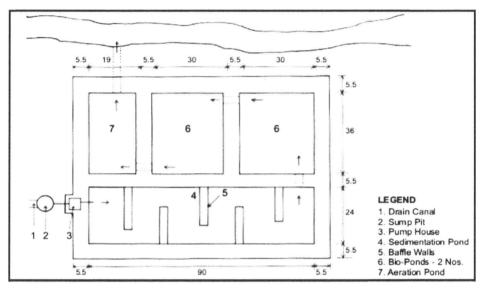

Fig. 24: Layout plan of effluent treatment system for a 5 hectare shrimp farm, CAA, 2002

Harvest

The supernatant water from the culture ponds up to a depth of 30 cm is drained starting from 10 days prior to the day of harvest.

$$\text{Volume of water/day} = 50,000 \times 0.7/10 = 3500 \text{ m}^3/\text{day}.$$

Although this supernatant water will carry practically little sediment load, a retention period of 15 hours, 11 hours and 8 hours will be required in the sedimentation, bio and aeration ponds respectively. The multiplication of 0.7 is used taking into consideration the combined seepage and evaporation loss (0.3). Evaporation of water from a water surface depends on water temp., air temp., air humidity and air velocity above the water surface.

Note

The Aquaculture Authority issued guidelines on the need for Effluent Treatment System (ETS) in shrimp farms. The guidelines state that all shrimp farms of 5 ha water spread area and above located within the CRZ, and 10 ha water spread area and above located outside the CRZ, should set up an ETS or effluent treatment ponds/facilities. The guidelines also refer to the need for a common ETS for clusters of shrimp farms, where each farm is lesser than 5 ha in size.

Table 37: Important steps of water and wastewater treatment for hatchery and culture practices

		Treatment step		
Sl. No.	Mechanical	Biological	Chemical: oxidation	Miscellaneous
1.	Screening	Aerobic	Chlorination	Activated carbon
2.	Sedimentation: coagulation/ liming	Anaerobic	Ozonation	UV radiation: the last treatment step
3.	Clarification and filtration	Facultative	Aeration	Specific problem control like pH, iron and ammonia *etc*.

CHAPTER 19

BIOWASTE AND ITS APPLICATION

INTRODUCTION

Wastes are materials that are not prime products (that is products produced for the market) for which the generator has no further use in terms of his/her own purposes of production, transformation or consumption. Wastes may be generated during the extraction of raw materials, the processing of raw materials into intermediate and final products, the consumption of final products and other human activities. Residuals recycled or reused at the place of generation are excluded.

Biowaste

Biowaste is a term used to describe organic waste that is putrescible - liable to decay or spoil. Biowaste is defined as animal and vegetative wastes arising from households, dairy, and the food processing and beverage industry. This can include food waste, biodegradable components of sewage, agricultural wastes and sludges. Biowaste means any waste that is capable of undergoing anaerobic or aerobic decomposition, such as food and garden waste, and paper and paperboard. There are two main sources of biowaste – municipal sources and industrial sources.

Municipal biowaste

Approximately two-third of the waste produced by household sources and commercial enterprises comprises 'organic' or natural materials. These materials get broken down over time ('biodegrade') by natural processes. This waste stream is termed as Biodegradable Municipal Waste (BMW). It comprises paper and cardboard, food waste, textiles, excreta (faecal matter) and wood. When land gets filled, these materials degrade and generate leachate and landfill gas.

BMW requires recycling or bio-treatment in order to avoid these problems and to keep away dependence on landfill as a disposal option.

Industrial biowaste

Industrial processes such as food processing, dairy, agriculture, forestry and pharma-chem industry are examples of processes that may produce large volumes of biodegradable waste streams. These materials are often highly putrescible and may be liquid in form. Therefore, bio-treatment is required to ensure that environmental protection can be assured.

Liquid biowaste

Liquid biowaste as the name envisages is biowaste in liquid state. Solid biowaste can also be converted to liquid biowaste, e.g.: cow dung being converted to cow dung slurry. Liquid biowastes are produced in huge quantities, like the case of sewage and domestic wastewater *etc.* Liquid biowastes also possess another advantage that, it being dominated by water, can be used simultaneously as a water source as well as the nutrient source.

Definition

- **'Municipal waste'** means waste from households, as well as other waste which, because of its nature or composition, is similar to waste from households.

- **'Compost'** means the stable, sanitised and humus-like material rich in organic matter and free from offensive odours resulting from the composting process of separately collected biowaste.

- **'Digestate'** means the material resulting from the anaerobic digestion of separately collected biowaste.

- **'Biogas'** means the mixture of carbon dioxide, methane and trace gases resulting from the controlled anaerobic digestion of biowaste.

- **'Stabilised biowaste'** means the waste resulting from the mechanical/biological treatment of unsorted waste or residual municipal waste as well as any other treated biowaste.

- **'Composting'** means the autothermic and thermophilic biological decomposition of separately collected biowaste in the presence of oxygen and under controlled conditions by the action of micro and macroorganisms in order to produce compost.

- **'Windrow composting'** means the composting of biowaste placed in elongated heaps which are periodically turned by mechanical means in order to increase the porosity of the heap and increase the homogeneity of the waste.

- **'In-vessel composting'** means the composting of biowaste in a closed reactor where the composting process is accelerated by an optimised air exchange, water content and temperature control.

- **'Home composting'** means the composting of the biowaste as well as the use of the compost in a garden belonging to a private household.

- **'On-site composting'** means the composting of the biowaste where it is generated.

- **'Community composting'** means the composting of biowaste by a group of people in a locality with the aim of composting their own and other people's biowaste in order to manage the supply biowaste as close as possible to the point at which it is produced.

- **'Anaerobic digestion'** means the biological decomposition of biowaste in the absence of oxygen and under controlled conditions by the action of microorganisms (including methanogenic bacteria) in order to produce biogas and digestate.

- **'Treatment'** means composting, anaerobic digestion, mechanical/chemical/biological treatment or any other process for sanitising biowaste.

- **'Mechanical/biological treatment'** means the treatment of residual municipal waste, unsorted waste or any other biowaste unfit for composting or anaerobic digestion in order to stabilise and reduce the volume of the waste.

- **'Separate collection'** means the collection of biowaste separately from other kinds of waste in such a way as to avoid the different waste fractions or waste components from being mixed, combined or contaminated with other potentially polluting wastes, products or materials.

- **'Residual municipal waste'** means the fraction of municipal waste remaining after the source separation of municipal waste fractions, such as food and garden waste, packaging, paper and paperboard, metals, glass, and unsuitable for the production of compost because it is mixed, combined or contaminated with potentially polluting product or material.

- **'Sanitation'** means the treatment of biowaste, during the production of compost and digestate that aims at killing organisms pathogenic to crops, animals and men, to a level that the risk of carrying disease in connection with further treatment, trade and use is minimised.

- **'Impurities'** means the presence of fragments of plastic, glass, metals or similar non-biodegradable materials, with the exclusion of sand, gravel and pebbles.

- 'Ecological improvement' means the maintenance of habitats and their biodiversity where these would otherwise deteriorate, the provision of new habitats for wild life and the development or restoration of existing habitats to give greater biodiversity and sustainability while ensuring the protection of the environmental quality in the broadest sense.

General principles of biowaste management

- An improved management of biowaste should be as per the following recommendations:

- The prevention or reduction of biowaste production (*e.g.,* sewage, sludge) and contamination by pollutants.

- The reuse of biowaste (*e.g.,* cardboard).

- The recycling of separately collected biowaste into the original material (*e.g.,* paper and cardboard) whenever environmentally justified.

- The composting or anaerobic digestion of separately collected biowaste, that is not recycled into the original material, with the utilisation of compost or digestate for agricultural benefit or ecological improvement.

- The mechanical/biological treatment of biowaste.

- The use of biowaste as a source for generating energy.

Treatment options for biowaste

Biowaste being biodegradable can be treated in many ways most of which make use of decomposition by natural means (bacteria *etc.*) of the biowaste into simpler and more useful forms with respect to the nature or environment. The biodegradable substances in sewage are also biowastes, however, having own treatment options. Even though 2/3rd of wastes in sewage are organic waste, 99% of sewage is water alone and thus it is dealt separately. Biowaste present in sewage has its own treatment schedules even though the basic processes underlining the operation are one and the same. This section mainly deals with the treatment options for the solid biowaste unlike liquid or smaller solid forms present in sewage. Solid biowaste would include a number of amplitudes from municipal biodegradable waste to industrial biodegradable wastes which when directly dumped in to the environment cause a number of nuisances to the public and hence need to be suitably treated. The following are the various treatment options.

(1) Composting

- On site composting
- Off site composting

(2) Home composting

(3) Anaerobic digestion

(4) Mechanical and biological treatment

(5) Community composting

(6) Separate collection

Negative impacts of biowaste accumulation in environment

Modernization and development have had their share of disadvantages and one of the main aspects of concern is the pollution, it is causing to the earth, *i.e.,* on land, air and water. With burgeoning global population and the rising demand for food and other essentials, there has been a rise in the amount of waste being generated daily by each household. This waste is ultimately thrown into municipal waste collection centres from where it is collected by the area municipalities to be further thrown into the landfills and dumps. However, either due to resource crunch or inefficient infrastructure, not all of this waste get collected and transported to the final dumpsites.

Preventive measures

Proper methods of waste disposal have to be undertaken to ensure that it does not affect the environment around the area or cause health hazards to the people living there. At the household-level, proper segregation of waste has to be done and it should be ensured that all organic matter is kept aside for composting, which is undoubtedly the best method for the correct disposal of this segment of the waste. In fact, the organic part of the waste that is generated gets decomposed more easily, attracts disease vectors and causes disease. Organic waste can be composted and then used as a fertilizer.

Conclusion

Biowaste is a misplaced source as well as a problem and proper treatment of biowaste would mean production of several products which can be used for many productive purposes. Biowaste recycling, production of fertilizers and production of energy (biogas) are some of the by-products of proper biowaste management and treatment. Overall, biowastes and wastewaters are misplaced sources which can be commercially very well exploited for aquaculture, aquatic health and human welfare through scientific and technical inputs; as aquaculture practices are quite efficient to convert biowastes/liquid biowastes into edible protein.

CHAPTER 20

WASTEWATER USE IN AQUACULTURE: SEWAGE FED AQUACULTURE

INTRODUCTION

Wastewater-fed aquaculture has a long history in several countries in East, South and South-East Asia. In China and other Asian countries (*e.g.,* Bangladesh, India, Indonesia and Vietnam), where the majority of global aquaculture production takes place, there is a long history of administering wastewater, animal waste and human excreta to fish ponds. Wastewater-fed aquaculture was mainly developed by farmers and local communities to increase food production.

Night soil, wastewater and polluted surface water were often used in aquaculture without any pretreatment. Systems as primarily engineered to treat wastewater that incorporated an aquaculture component appeared latter. About 90 systems were constructed in Germany between the end of the 19th century and the 1950s. Sewage-fed fish ponds were developed later in Asia than in Europe. They appeared in India in the 1930s, in China from the 1950s onwards and in Vietnam in the 1960s.

Sewage characteristics and treatment

The quantity of water consumption ranges approximately from 60-150 litres per capita per day. On an average, 9 g of nitrogen and 2 g of phosphorus per person per day are generated as waste. International norms put BOD generation at 0.1 kg per person per day. Sewage contains N, P and K concentrations as follows:

Nitrogen (N)	-	10-70 mg/l
Phosphorus (P)	-	7-20 mg/l
Potassium (K)	-	12-30 mg/l

Sewage needs the treatment to control following parameters so as to make it safe before taking to sewage fed aquaculture.

1. Higher quantity of obnoxious gases like ammonia, carbon dioxide, hydrogen sulphide and methane.

2. Higher concentration of pathogens.

3. Higher BOD load and organic carbon value.

4. Higher amount of suspended matter.

5. Very low dissolved oxygen content.

Food chains

The objective to fertilize an aquaculture pond with excreta, night soil or liquid biowastes is to produce natural food for fish. Since several species of fish feed directly on faecal matter, use of raw sewage or fresh night soil as influent to fish ponds should be prohibited for health reasons. Edwards (1986) has represented the complex food chains in an excreta-fed fish pond as shown in the figure no. 25, involving ultimate decomposers or bacteria, phytoplankton, zooplankton and invertebrate deteretivores. Inorganic nutrients released in the bacterial degradation of organic solids in sewage, night soil or excreta are taken up by phytoplankton. Zooplankton graze phytoplankton and small detritus particles coated with bacteria, the latter also serve as food for benthic invertebrate deteretivores. Plankton, particularly phytoplankton, are the major sources of natural food in a fish pond but benthic invertebrates, mainly chironomids, also serve as fish food, although they are quantitatively less important. To optimize fish production in a human waste fed pond, the majority of the fish should be filter feeders, to exploit the plankton growth.

Fish species

A wide range of fish species has been cultivated in aquaculture ponds receiving sewage, including common carp *(Cyprinus carpio)*, Indian major carps *(Catla catla, Cirrhinus mrigala and Labeo rohita)*, silver carp *(Hypophthalmichthys molitrix)*, bighead carp *(Hypophthalmichthys nobilis)*, grass carp *(Ctenopharyngodon idella)*, crucian carp *(Carassius carassius)*, Nile carp *(Osteochilus hasseltii)*, tilapia *(Oreochromis spp.)*, milkfish *(Chanos chanos)*, catfish *(Pangasius spp.)*, kissing gouramy *(Helostoma temmincki)*, giant gourami *(Osphronemus goramy)*, silver barb *(Puntius gonionotus)* and freshwater prawn *(Macrobrachium rosenbergii)*.

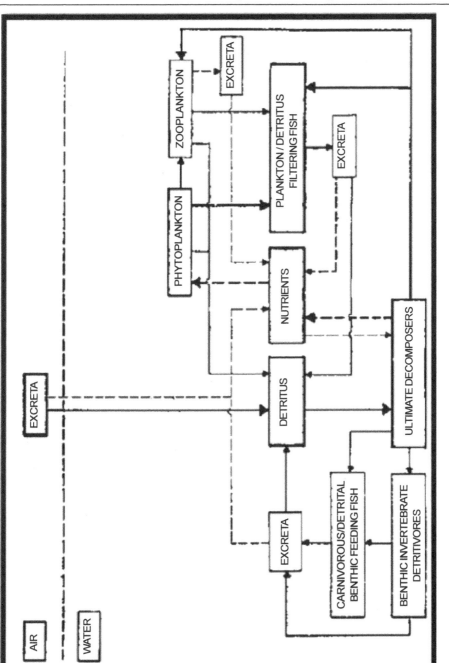

Fig. 25: Food chain in sewage fed aquaculture as given by Edwards et al., 1986

The selection reflects local culture rather than fish optimally-suited to such environments. For example, Chinese and Indian major carps are the major species in sewage-fed systems in China and India respectively. In some countries, a polyculture of several fish species is used. Tilapia are generally cultured to a lesser extent than carps in sewage-fed systems although, technically, they are more suitable for this environment because they are better able to tolerate adverse environmental conditions than carp species. Milkfish have been found to have poorer growth and survival statistics compared with Indian major carps and Chinese carps in ponds fed with stabilization pond effluent in India.

Although fish are generally divided into types according to their natural nutritional habits - those that feed on phytoplankton, zooplankton and benthic animals - several species are known to feed on whatever particles are suspended in the water. There is also uncertainty about the types of phytoplankton fed upon by filter-feeding fish. For example, although blue-green algae are thought to be indigestible to fish, tilapia has been seen to readily digest these algae and there is evidence that silver carp can do the same. Considering the high carrying capacity and more productivity of sewage-fed ponds, fish stocking can be done at reasonably higher density, ranging from 10,000 to 15,000 fingerlings/ha, including omnivorous, scavengers and bottom feeders.

Aquatic plants

Aquatic macrophytes grow readily in ponds fed with human waste and their use in wastewater treatment has already been established. Some creeping aquatic macrophytes are cultivated as vegetables for human consumption in aquaculture ponds and duckweeds are also cultivated, mainly for fish feed. Among the aquatic plants grown for use as vegetables are water spinach *(Ipomoea aquatica),* water mimosa *(Neptunia oleracea),* water cress *(Rorippa nasturtium-aquaticum)* and Chinese water chestnut *(Eleocharis dulcis).* The duckweeds like *Azolla, Lemna, Spirodela* and *Wolffia* are cultivated in some parts of Asia in shallow ponds fertilized with excreta, mainly as feed not only for Chinese carp but also for chickens, ducks and edible snails (Edwards, 1990). Duck weeds in sewage-fed ponds can grow at the following rate.

Azolla: 160 g/m³/d

Lemna: 275 g/m³/d

Spirodela: 350 g/m³/d

Wolffia: 280g/m³/d

Fundamental principles in wastewater-fed aquaculture

Wastewater-fed aquaculture has been practiced throughout the world. Aquaculture and wastewater treatment appear to be two independent and parallel aspects. Most traditional wastewater reuse systems were designed to produce fish from nutrients available in wastewater. In contrast, conventional sewerage systems such as activated sludge, trickling filters and stabilization ponds are designed to treat wastewater and thereby remove nutrients. Actually, both are connected and complementary and are parts of the entire function of an aquatic ecosystem. According to the ancient Chinese philosophy from "the Book of Changes", it is the universal law 'to move in an endless cycle, go round and begin again'.

This thinking is of momentous current significance in seeking the path towards sustainable development and betterment of human society. Its fundamental principles are regeneration and circulation. Wastewater-fed aquaculture systems provide high fish yields at a low cost since little to no additional inputs, such as formulated feeds are administered to the fish. The waste is consumed directly by the fish and also provides nutrients for the growth of photosynthetic organisms. Fish and other aquatic organisms, raised in wastewater-fed systems provide an important source of food and employment for millions of people. Conversion of waste into usable resources, environmental protection, internalized and sustainable development; all are realized in the wastewater-fed aquaculture system. The regeneration and circulation of substances are exactly the fundamental strategies for the survival and development of natural ecosystems.

Table 38: Types of wastewater-fed aquaculture systems

Wastewater type	Aquaculture system	Culture organism	Country
Night soil (overhanging latrine)	Pond	Fish, Duckweed	China, Indonesia, Vietnam, Bangladesh
Night soil, septage (cartage)	Pond	Fish, Duckweed	China, Vietnam
Contaminated surface water (waterborne)	Pond	Fish, Duckweed	Bangladesh, Indonesia, Vietnam, China
Contaminated surface water (waterborne)	Cage in river	Fish	Indonesia
Seepage (waterborne)	Pond	Fish, Duckweed	China, India, Germany, Vietnam, Bangladesh

Table 39: Characteristics of liquid biowastes

Sl. No.	Source	ppm					Algal load
		BOD*	COD	TSS	N	P	
1.	Sewage	100-400	200-700	300-1000	10-70	7-20	95-380
2.	Dairy	800-2500	1200-3750	100-1000	48-150	10-100	760-2375
3.	Sugar	400-1500	1500-4000	300-700	10-15	15-60	380-1425
4.	Beverage	40000-50000	80000-90000	5000-7000	500-600	5-7	38000-47500
5.	Confectionery	3000-7000	6000-10000	900-1200	50-75	2-4	2850-6650
6.	Distillery	30000-35000	50000-60000	4500-5500	1200-1400	44-55	28500-33250

*The algal load or natural fish food that can be generated from BOD load is = BOD x 0.95; thus the higher BOD loads of different liquid biowastes have comparatively the greater nutritional strength to accelerate aquatic fertility and productivity.

Health issues and waste-fed aquaculture

Based on recommendations by WHO (1989) and bacterial quality standards and threshold concentrations for fish muscle, Pullin *et al.* (1992) published guidelines for domestic wastewater reuse in aquaculture:

- A minimum retention time of 8-10 days for raw sewage.

- A tentative maximum critical density of 105 total bacteria/ml in wastewater-fed fish pond water.

- Absence of viable trematode eggs in fish ponds.

- Suspension of wastewater loading for 2 weeks prior to fish harvest.

- Holding fish for a few hours to facilitate evacuation of gut contents.

- The < 50 total bacteria/g of fish muscle and no Salmonella.

- Good hygiene in handling and processing, including evisceration, washing and cooking.

- Use as high-protein animal feed if direct consumption of fish is socially unacceptable.

CHAPTER 21

CAUSES OF SOIL SALINIZATION

1. SALINITY AND ALKALINITY

Saline aquaculture is a scientific tool to solve a serious environmental problem of soil salinization. As soil is not a renewable source, therefore, it needs the measures like protection, amendment and improvement for its better utilization. Soil salinity is the salt content present in soil; while the process of getting salt content enhanced accumulation is known as soil salinization. The salt-affected soils are those where salt levels reduce agriculture yield. Such soils have the saline subsoil water which can be very well used for saline aquaculture. The water and soil unfit for human consumption and agriculture purpose respectively can be very well exploited for aquaculture. Salinity and alkalinity are two different terms and therefore, saline (sodium dominance) and alkaline (carbonate dominance) soils also differ from each other. Knowledge of this difference is essential to chalk out a correct reclamation programme. The difference lies in the type of salt contents present in the soil. Saline soil contains a very high concentration of calcium chloride, magnesium chloride, calcium carbonate and magnesium carbonate; whereas alkaline soil holds a high concentration of one salt called sodium carbonate. Other salts of calcium, magnesium and sodium with chloride, sulphate and carbonate may also be there in the soil. In alkaline soil, concentration of sodium is too high. Gypsum (calcium sulphate) which is the most widely used amendment for salt affected soils should be used in alkali soils. Using gypsum in saline soil may even aggravate the problem. Based on the type and concentration of salts, salt affected soils can be classified into following three groups.

A. Saline soils

As discussed above, soils containing excess amount of water soluble salts are called saline soils. The concentration of salt in the root-zone of soil goes so high that it adversely affects plant growth. Looking from a distant place, these soils give a milky white or whitish-grey appearance. Therefore, sometimes these soils are also referred to as 'white alkali soils'. Salt concentration of these soils is expressed in terms of electrical conductivity (unit: mmhos/cm). Electrical conductivity of saline soils is always more than 4 millimhos/cm, pH (soil reaction) is more than 7 but always lesser than 8.5. It is feared that introduction of new irrigation projects, canal irrigation system *etc.* may lead to increase in salinity problem, especially if proper soil management and arrangement of drainage system is not made.

B. Alkaline soils

A number of elements in their ionic form, *i.e.*, extremely minute electrically charged particles remain adhered to soil particles. These elements are exchangeable with other elements found in soil water. When the concentration of sodium (Na) in the soil goes so high that it occupies more than 15% of the total exchangeable site surface of soil particles, soil is said to be alkaline and sodium in this state is called exchangeable sodium percentage (ESP). Concentration of salts other than sodium carbonate is low in this soil. Therefore, electrical conductivity is lower than 4 millimhos/cm and the pH of the soil is more than 8.5. Calcium carbonate, mixed with soil water and organic matter, imparts black colour to the soil. Subsoil of alkali soils often contains a hard pan which inhibits the downward movement of water.

C. Saline-alkali soils

These soils contain all types of salts including sodium salt. Electrical conductivity of these soils is more than 4 millimhos/cm and pH more than 8.5. They also develop a hard pan in sub-soil. They are grey in colour. A visual diagnosis of salt affected soils is possible by looking at the colour of the soil. Completely white soils are saline, grey are saline-alkali and black are alkali soils. Presence of hard pan is also a distinguishing character of alkaline soils.

Table 40: Characteristics of saline, alkaline and saline-alkaline soils

Sl. No.	Soil	Electrical conductivity (mmhos/cm)	Exchangeable sodium %	pH
1.	Saline	> 4	< 15	< 8.5
2.	Alkali	< 4	> 15	> 8.5
3.	Saline-alkali	> 4	> 15	> 8.5

2. CAUSES OF SALT ACCUMULATION

A number of factors are responsible for the formation of saline and alkaline soils. These factors are mainly geological, climatic and hydrological in nature and are discussed below.

A. Geological factors

Geological factors relate to the origin of the soils. Some soils are formed from basic parent material (rocks-basalt and gabbro *etc.*); therefore, saline in nature. Some soils contain salt layers in sub-soil. When such soils are deep cultivated, these salts come up to the surface soil. Some saline soils are also believed to have been formed by deposition of salts carried down by rivers from hill rocks. Soils of coastal zones, which were once inundated by sea water and now are within the reach of sea breeze, are also saline.

B. Climatic factors

Soluble salts accumulate whenever evaporation exceeds total precipitation (rain) either alone or in combination with irrigation. Total evaporation in arid zones reaches 1500-3000 mm per year, which often exceeds the total precipitation received even for a number of years.

C. Hydrological factors

Water is the chief transporting agent for salts and its evaporation gives rise to accumulation of salts in the soil. At flood plains, deltas, coastal areas, lakes and regions of high water table, surface movement of water is negligible. Water, therefore, evaporates leaving the salts on the surface. This process also brings up salts from the lower horizons, if the moisture regime is connected with ground water. In fact, this is the most important cause of salinization.

If the depth of water-table below the soil surface is lesser than 120 cm, salinization of surface soil may occur. Ascending water table towards the surface dissolves salts falling in the way and leaves them on the soil surface after evaporation. Movement of water from lower soil layers to surface layers takes place through capillaries (narrow tubular water path ways formed by the arrangement of soil particles). Chances of salinization increase if the ground water is salty (salt content more than 2 g/litre). The arid and semi-arid areas have often poor quality ground water. The use of saline ground water for irrigation has accentuated the problem of salt accumulation in several areas. It is not rare to find such examples at many places of Punjab, Haryana, Rajasthan, Western U.P. and Gujarat in India having shallow depth of utilizable saline ground water which has already created problems of soil salinity. The problem is further aggravated where the ground water has high sodium content. The fact that the introduction of canal irrigation system has been the principal cause of extension of salt in many parts is now well established. This is not because of canal water is salty but because water table of the surrounding soils is increased and the native salts are pushed up towards soil surface. Removal of plant cover also enhances evaporation of water from soil surface and, therefore, salinization gets increased.

3. SALINITY MEASUREMENT AND UNIT CONVERSION

Soil and water salinity is often measured by electrical conductivity (EC) and commonly used EC units are deciSiemens per meter (dS/m) and millimhos per centimeter (mmhos/cm). Numerical relationships in between EC and TDS are as follows:

1 dS/m = 1 mmhos/cm

TDS (ppm or mg/l) = EC (dS/m) x 640 if EC is 0.1 to 5 dS/m.

TDS (ppm or mg/l) = EC (dS/m) x 800 if EC is >5.0 dS/m.

In brackish water aquaculture and mariculture practices, the measurement unit often used for water salinity is ppt. One ppt water salinity is equivalent to 1g/litre or 1000 mg/litre or 1000 ppm of total dissolved solids (TDS).

Note

The most important differences in the waters of inland saline aquaculture and common coastal aquaculture are ionic composition, concentration and the relative available ratio of ions. Among these three factors, calcium and magnesium ionic ratio is very crucial for successful inland saline aquaculture and this ratio can be as, Ca: Mg = 1: 3.375. This is because; the elemental concentrations of Ca and Mg in the sea water at 35 ppt salinity are 400 mg/l and 1350 mg/l respectively, *i.e.*, 1: 3.375. This approximate 1: 4 Ca and Mg ratio is very important for osmoregulation, moulting, shell formation and physiological functions. The imbalance among ions' concentration and ratio will lead to osmotic stress, in turn affecting growth, survival and productivity. If necessary, the mineral supplementation can be made in inland saline water based on desired levels of ionic concentration and ratio. These desired levels at different salinity dilutions can be ascertained proportionately based on following bench mark elemental concentrations (ppm) at 35 ppt salinity.

K = 380; Ca = 400; SO_4 – S = 885; Mg = 1350; Na; 10500; Cl = 19000.

CHAPTER 22

PLANNING AND MANAGEMENT OF RESERVOIR FISHERIES

INTRODUCTION

Planning and management are relatively new subjects. Planning is for tomorrow and management is for today. These subjects have acquired greater importance during the past two decades. The purpose of planning is (1) to match the limited resources with many problems, (2) to eliminate wasteful expenditure or duplication of expenditure, and (3) to develop the best course of action to accomplish a defined objective. The increasing demand for aquaculture and environmental health care activities, in the face of limited resources, have brought out the need for careful planning and management of aquatic resources. Planning and management are considered essential if higher standards of environmental health care are to be achieved. Planning is a matter of team work and consultation. The planning team consists of not only specialists within the field, but also specialized in other fields, *viz.,* economics, statistics, sociology and management *etc.* Planning in its broadest sense includes three steps:

(a) Plan formulation

(b) Execution

(c) Evaluation

Development planning

Every country has its own plan for national development. The purpose of national planning is to achieve a rapid, balanced, economic and social development for the country as a whole. The National Development Plan of a country is a combination of sectorial plans which comprise the following sectors, *viz.,* food and agriculture, education, health and family planning, industry, transport and communications, housing, power and social welfare *etc*. All these sectors compete for national resources. In this context, National Development Planning has been defined as, "continuous, systematic, coordinated, and planning for the investment of the resources of a country (men, money and material) in programmes aimed at achieving the most rapid economic and social development possible". But successful planning also needs reserves of healthy aquatic environment. Therefore, water and wastewater management is very much instrumental to further promote fisheries and aquaculture.

Reservoir planning

The tropical reservoirs are primarily constructed for irrigation, hydroelectricity generation and water supply schemes taking into consideration their catchment and command areas. Inland fisheries development is a secondary use of most reservoirs. When barrier is constructed across some river in the form of a dam, water gets stored up on the upstream side of the barrier, forming a pool of water generally called a reservoir. Broadly speaking, any water collected in a pool or a lake may be termed as a reservoir. The water stored in a reservoir may be used for various purposes. Depending upon the purposes served, the reservoirs may be classified as follows:

(1) Storage or Conservation Reservoirs

(2) Flood Control Reservoirs

(3) Distribution Reservoirs

(4) Multipurpose Reservoirs

Enhancement of reservoir productivity

How to enhance reservoirs' primary aquatic productivity whose waters are used for human consumption?

One cropping of *Sesbania* spp. in a year, on the peripheral area of the reservoir, and when the crop is ready within 2-3 months duration after sowing; its mixing in the soil of reservoir peripheral area will not only enhance the aquatic primary productivity 2-3 times within a 3 to 4 years span of *Sesbania* spp. cropping but will also improve long term edaphic nutrients availability status (Table 41).

Table 41: Observations before and after *Sesbania aculeata* plantation

Sl. No.	Parameter	Before plantation	After plantation
WATER			
1.	pH	6.7 – 6.9	7.3 – 7.5
2.	Alkalinity: mg/l	35 - 40	55 - 60
3.	Hardness: mg/l	33 - 37	50 - 55
4.	NO_3 - N: mg/l	0.05 – 0.07	0.10 – 0.20
5.	PO_4 - P: mg/l	0.01 – 0.02	0.05 – 0.07
6.	Aquatic primary productivity: mg C m^{-3} day^{-1}	235 - 250	480 - 550
SEDIMENT			
7.	pH	5.6– 5.8	6.0 – 6.5
8.	Organic carbon (%)	0.20 – 0.25	0.50 – 0.80
9.	Nitrogen (mg/100 g)	25 - 30	50 - 67
10.	Phosphorus (mg/100 g)	2.8 – 3.0	6.0 – 6.4

Stocking in cages

- Generally a minimum stocking density of 100 fry/m^3 is recommended.
- Desired production at harvest 150 kg/m^3.
- Desired average number of fish at harvest:

$$\text{No. of stock} = \frac{\text{Total fish weight at harvest } (150\text{kg/m}^3)}{0.5 \text{ kg (desired average weight of one fish at harvest)}}$$

$$= 300 \text{ fry/m}^3$$

Fig. 26: *Sesbania aculeata*

CHAPTER 23

NUTRIENTS BUDGETING FOR FISH CULTURE

INTRODUCTION

Application of manures and fertilizers in aquaculture ponds is done to enhance the carrying capacity of the system in turn to increase the production per unit of area/volume. Further, by the application of fertilizers and manures, nutrient status of the pond water is enhanced which accelerates the natural food production required for the fishes being cultured. Manures and fertilizers mainly used in aquaculture practices are raw cattle dung, pig dung, poultry litter; and urea, single super phosphate and muriate of potash respectively; their budgetary requirement is depicted in the tables below.

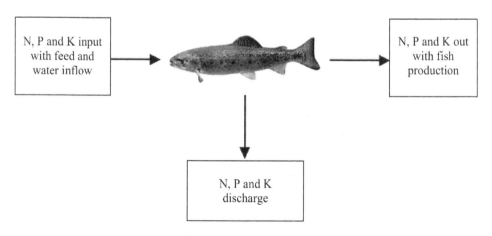

Fig. 27: Nutrients flow

Table 42: NPK status of organic manure

Sl. No.	Name	Percent (%)			Organic carbon
		N	P	K	
1.	RCD	0.3	0.2	0.1	8.0 – 10.0
2.	Pig dung	0.6	0.5	0.4	8.0 – 9.0
3.	Poultry litter	1.4	0.8	0.6	5.0 – 7.0

Table 43: Nutrient budgeting for aquaculture using inorganic fertilizers

Sl.No.	Mineral element		Ratio		Concentration					(mg/l)	Fertilizer requirement	
	Name	Chemical symbol	Available in natural sources	Desirable for aquaculture	Available in natural sources (mg/l)	Desirable for aquaculture (mg/l)	Fertilizer supplementation				(Kg/h) (elemental form)	(kg/h) (fertilizer form)
							Name	Chemical formula	Mineral availability (%)			
1.	Nitrogen	N	25	4	> 0.25	0.50 - 4.5	Urea	$NH_2 - CO - NH_2$	46	2.0	20	43.5@ 46% available in urea
2.	Phosphorus	P	1	1	> 0.01	0.05 - 0.40	Single super phosphate	$CaH_3(PO_4)_2.H_2O$ (30%), $CaHPO_4$ (10%), $CaSO_4$ (45%), iron oxide, alumina and silica (10%), water (5%)	16	0.25	2.5	15.5@ 16% available in single super phosphate
3.	Potassium	K	50	3	> 0.50	0.50-2.0	Muriate of potash	KCl	52	0.75	7.5	14.5 @ 52% available in muriate of potash

Table 44: Nutrient budgeting for aquaculture using organic manures

Organic manure	Nutrient symbol	Nutrient in organic manure (%)	Nutrient requirement for aquaculture pond (ppm)	Nutrient requirement in elemental form (Kg/ha)	Requirement of manure (Kg/ha)
Raw cattle dung (RCD)	N	0.3	2.0	20	30 × 100/0.6
	P	0.2	0.25	2.5	
	K	0.1	0.75	7.5	
Total		**0.6**		**30**	**=5000**
Pig manure	N	0.6	2.0	20	30 × 100/1.5
	P	0.5	0.25	2.5	
	K	0.4	0.75	7.5	
Total		**1.5**		**30**	**=2000**
Poultry manure	N	1.4	2.0	20	30 × 100/2.8
	P	0.8	0.25	2.5	
	K	0.6	0.75	7.5	
Total		**2.8**		**30**	**=1071.43**

CHAPTER 24

AQUATIC PRODUCTIVITY ENHANCEMENT THROUGH NUTRIENTS MANAGEMENT

INTRODUCTION

To augment the world food supply, efforts are being made to obtain higher production of commercial important species and search for new candidate species. As the land resources are not sufficient to feed the growing population, therefore, vertical and horizontal expansions of aquaculture are essential to produce cheaper protein sources from fallow land and water bodies. In this regard, the burgeoning science of aquaculture has to play an important role for the enhancement of aquatic productivity. In this chapter, efforts have been put on the status of nutrients and their budgeting in aquculture system. These nutrients are the specific substances needed for the growth of organisms. Various nutrients are required for photosynthetic activity, which in turn, enhance the pond aquatic productivity. Fish production depends on relative quantum of nutritive substances available in ponds. These aspects can be practically made possible after application and management of nutrients doses under their favourable limits, having the knowledge of status of nutrients and planning for their future budgetary requirement.

Nutrients are the chemical substances essentially required for the growth of organisms. They are of three types like primary nutrients such as nitrogen, phosphorus and potassium; secondary nutrients, *i.e.*, calcium, magnesium, silica and sulphur and last but not of least importance like iron, copper, zinc, cobalt, manganese and iodine *etc.*, they are tertiary nutrients and also known as micronutrients or trace elements. Nutrients' budgeting plays a very important role in the fish production enhancement aspect for aquaculture management practices.

Their proper monitoring and budgeting amplitudes can burgeon aquatic body status from atrophic to desired trophic level.

Table 45: Classification of water bodies on the basis of nutrients availability

Sl. No.	Type of water body	*Nature of aquatic medium	pH range	Productivity level (mg C/m³/ day)
1.	Atrophic	Acidic	< 6.8	Almost nil
2.	Oligotrophic	Slightly acidic to a little alkaline	6.9-7.5	50-500
3.	Mesotrophic	Moderately alkaline	7.5-8.5	500-1000
4.	Eutrophic	Highly alkaline	8.5-9.0	1000-2500
5.	Highly eutrophic	Hyper alkaline	9.0-9.5	> 2500

*Acidic waters make the nature of medium corrosive, on the other hand, highly alkaline conditions turn it in caustic nature and thus both limit the fish production.

Nutrients budgeting and calculation

To make the calculations for nutrient budgeting their desired levels of concentrations can be known from the table number 43 and 44 as the bench mark; while here, the four following examples are being given for nitrogen, phosphorus, potassium and sediment organic carbon respectively.

Example 1: Nitrogen budgeting supplemented by urea

- The calculation is based on one hectare water surface area, average depth 1 meter.

- Suppose, the nitrogen availability in the system is 0.5 ppm.

- Suppose, the desired level is 3.0 ppm.

- The demand required is 3.0 - 0.5 = 2.5 ppm or 2.5 mg/l.

- The available nitrogen in urea is 46%.

- The demand for one hectare water surface area is: (Water volume: 1m × 10,000 m²)

$$= \frac{1m \times 10,000 \ m^2 \times 1000 \times 2.5 \times 100}{46 \times 1000 \times 1000}$$

$$= \frac{10 \times 2.5 \times 100}{46}$$

$$= 54.347 \ kg$$

Example 2: Phosphorus budgeting supplemented by single super phosphate

- The calculation is based on one hectare water surface area, average depth 1 meter.
- Suppose the phosphorus availability in the system is 0.05 ppm.
- Suppose the desired level is 0.50 ppm.
- The demand required is 0.50 - 0.05 = 0.45 ppm or 0.45 mg/l.
- The available phosphorus in single superphosphate is 16%.

The demand for one hectare water surface area is: (Water volume: 1m × 10,000 m²)

$$= \frac{1m \times 10,000 \text{ m}^2 \times 1000 \times 0.45 \times 100}{16 \times 1000 \times 1000}$$

$$= \frac{10 \times 0.45 \times 100}{16}$$

$$= 28.125 \text{ kg.}$$

Example3: Potassium budgeting supplemented by muriate of potash

- The calculation is based on one hectare water surface area, average depth 1 meter.
- Suppose the potash availability in the system is 0.50 ppm.
- Suppose the desired level is 2.0 ppm.
- The demand required is 2.0 - 0.50 = 1.50 ppm or 1.50 mg/l.
- The available potassium in muriate of potash is 52%
- The demand for one hectare water surface area is: (Water volume: 1m × 10,000 m²)

$$= \frac{\text{Volume }(m \times m^2) \times (m^3 \text{ to litres}) \times \text{demand} \times \%}{\text{mg to kg}}$$

$$= \frac{1m \times 10,000 \text{ m}^2 \times 1000 \times 1.5 \times 100}{52 \times 1000 \times 1000}$$

$$= \frac{10 \times 1.5 \times 100}{52}$$

$$= 28.846 \text{ kg.}$$

Example 4: Sediment organic carbon budgeting supplemented by RCD

- The calculation is based on one m^2 of bottom surface area.
- Suppose the organic carbon availability in RCD is 8%
- *Suppose the organic carbon in the sediment is 0.5%
- The desired level is 2.0%
- The demand required is 2.0 - 0.5 = 1.5%
- Sediment depth taken for calculation is 8 cm.
- The demand for 1 m^2 bottom surface area is:

$$= \frac{16500 \times 1.5 \times 100}{1000 \times 100 \times 8}$$
$$= \frac{16.5 \times 1.5}{8}$$
$$= 3.093 \text{ kg.}$$

*Particle density of soil is 1.65 g/ml, while bulk density is 2.5 g/ml.

Aquatic productivity

Fill two light bottles and one dark bottle with water sample. Measure the initial DO concentration of one of the light bottles. Suspend the remaining two, one light and one dark bottles with the help of plastic rope at the desired depth, exactly at the depth from where the water samples are drawn. Let the incubation continue, the ideal photoperiod is sun rise to sun set. Estimate the DO concentration in light and dark bottles.

$$\text{Gross Primary Productivity} = \frac{LB - DB \times 0.375}{T} \times \frac{1000 \, \text{mg} \, Cm^{-3} \, \text{hour}^{-1}}{PQ}$$

$$\text{Net Primary Productivity} = \frac{LB - IB \times 0.375}{T} \times \frac{1000 \, \text{mg} \, Cm^{-3} \, \text{hour}^{-1}}{PQ}$$

Where

LB: Dissolved oxygen in light bottle

DB: Dissolved oxygen in dark bottle

IB: Dissolved oxygen in initial light bottle

T: Incubation period (in hrs)

C: Carbon synthesised during photosynthesis

PQ: Photosynthetic quotient = 1.2 (CO_2 intake ÷ O_2 output)

0.375 is C/O_2 = 12/32 = 0.375 (conversion of O_2 analysed to C synthesised)

1000 is multiplication of O_2 mg/L to m^3)

The tertiary productivity (expected fish production) can be calculated by the following formula:

Net Primary Productivity (mgC m^{-3} hr^{-1}) × 7.47 = NPP (mg m^{-3} hr^{-1}) as wet weight

Wet weight/100 = Tertiary productivity (mg m^{-3} hr^{-1})

$$\text{Tertiary productivity (Kg ha}^{-1}\text{ yr}^{-1}) = \frac{\text{mg m}^{-3}\text{hr}^{-1} \times 12 \times 365 \times 10^4}{10^6}$$

Where 12, hours of sun shine (sun rise to sun set)

365, number of days in a year

10^4, conversion of m^3 or m^2 to hectare

10^6, conversion of mg to kg

DETERMINATION OF BIOMASS

The desirable aquatic primary productivity will generate better growth and higher production of phytoplankton. Their biomass can be quantified as follows:

1. Phytoplankton standing crop method

Biomass is a quantitative estimate of the total mass of living organisms within a given area or volume. It may include the mass of a population or of a community but gives no information on community structure or function. The most accurate methods for estimation of biomass are dry weight, ash free dry weight and volume of living organisms.

100 – 125 mg carbon = 1ml or 1 cc phytoplankton.

2. Chlorophyll – a method

It is used as an algal biomass indicator assuming that chlorophyll – a, which constitutes on an average 1.5% of the dry weight of organic matter (ash free weight) of algae. The algal biomass estimation is chlorophyll – a content x 67.

3. Biochemical oxygen demand method

Algal biomass (primary productivity) = BOD × 0.95

Plankton count

Filter known quantity of water sample through a plankton net of bolting silk no. 25. Transfer the water sample into centrifugal tube and add a few drops of 5% formalin. Centrifuge the sample for 15 minutes. Put the centrifuged plankton in Sedgwick-Rafter counting cell and make the strip counting. The quantitative estimation of the plankton is done with a formula given below.

$$N = (a \times 1000) \times c/L$$

Where:

N = Number of plankton per litre of water sample

a = Average no. of plankton in 1 mm^3 capacity (plankton average no. in one small square count (1 mm^3 capacity) among 1000 squares (1 ml or 1000 mm^3 cavity capacity, the whole capacity of counting cell)

c = Volume of plankton concentration filtered in ml

L = Volume of sample water filtered in litres.

Plankton preservation

Zooplankton: 3 to 5% formalin; **Phytoplankton:** Lugol's solution.

To preserve samples with Lugol's solution add 0.3 ml Lugol's solution to 100 ml sample and store in dark place. For long term storage, add 0.7 ml Lugol's solution per 100 ml sample after 1hr. Prepare Lugol's solution by dissolving 20 g potassium iodide (KI) and 10 g iodine crystals in 200 ml distilled water containing 20 ml glacial acetic acid.

Table 46: Biochemical reactions affecting pH in water

Sl. No.	Process	Reaction	Effect on pH
1.	Photosynthesis	$6CO_2 + 6H_2O = C_6H_{12}O_6 + 6O_2$	Increase
2.	Respiration	$C_6H_{12}O_6 + 6O_2 = 6CO_2 + 6H_2O$	Decrease
3.	Nitrification	$NH_4^+ + 2O_2 = NO_3 + H_2O + 2H^+$	Decrease
4.	Denitrification	$5C_6H_{12}O_6 + 24NO_3 + 24H+ = 30CO_2 + 12N_2 + 42H_2O$	Increase
5.	Sulphide oxidation	$HS^- + 2O_2 = SO_4^{2-} + H^+$	Decrease
6.	Sulphate reduction	$C_6H_{12}O_6 + 3SO_4^{2-} + 3H^+ = 6CO_2 + 3HS + 6H_2O$	Increase
7.	Methane fermentation	$C_6H_{12}O_6 + 3CO_2 = 3CH_4 + 6CO_2$	Decrease

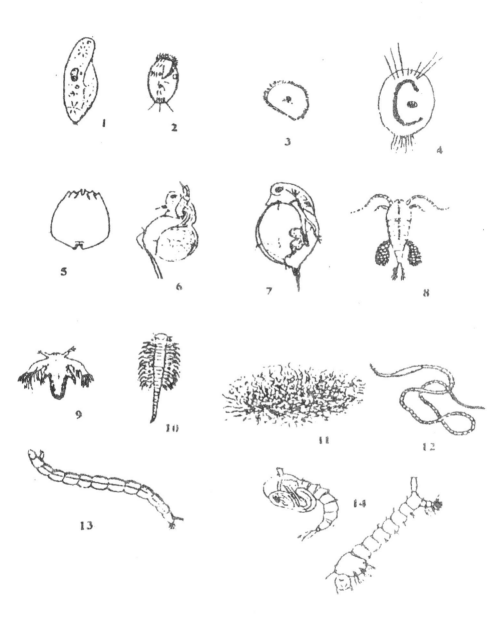

Fig. 28: Common zooplankton: 1.Paramecium, 2. Stolonychia, 3. *Fabria salina,* 4. Euplotes, 5. Brachionus, 6. Moina, 7. Daphnia, 8. Cyclops, 9. Artemia nauplius, 10. Adult Artemia, 11. A cluster of Tubifex worms, 12. Tubifex worm, 13. Chironomous larva, 14. Mosquito larvae

CHAPTER 25

ENVIRONMENTAL STRESS AND DISEASE CONTROL

INTRODUCTION

During recent years, there has been a worldwide upsurge of interest in maintaining fresh water and marine fishes in confined waters. Aquarists have been generally interested from decoration and exhibition points of view. The accelerating use of aquatic animals in research and burgeoning science of aquaculture are propagating the marketable aquatic species production on a commercial scale. In any case, the successful and economically feasible results depend on the management of stable water condition as a healthy environment is necessary for better growth and higher survival. The optimum or desirable water quality parameters in aquarium and garden pools to prevent occurrence of diseases are as follows:

Table 47: Optimum values of water quality parameters for ornamental fishery

Sl. No.	Parameter	Value
1.	Turbidity (NTU)	Nil
2.	Suspended solids (mgl^{-1})	30-50
3.	Temperature (°C)	
	a. Tropical climate	25-32
	b. Temperate climate	10-12
4.	pH	
	a. Fresh water	6.7-9.0
	b. Saline water	8.0-8.5
5.	Alkalinity (mgl^{-1})	50-175

[Table Contd.

Contd. Table]

Sl. No.	Parameter	Value
6.	Acidity (mgl^{-1})	< 50
7.	Hardness – fresh water (mgl^{-1})	50-175
8.	Salinity – saline water (ppt)	30-37
9.	Dissolved oxygen (mgl^{-1})	5-10
10.	Dissolved free carbon dioxide (mgl^{-1})	< 3
11.	Total nitrogen (mgl^{-1})	0.5-4.5
12.	Ammonia nitrogen (mgl^{-1})	
	a. Unionized (NH_3-N)	0-0.1
	b. Ionized (NH_4^+-N)	0-1.0
13.	Nitrite nitrogen (NO_2-N: mgl^{-1})	0-0.5
14.	Nitrate nitrogen (NO_3-N: mgl^{-1})	0.1-3.0
15.	Total phosphorus (mgl^{-1})	0.05-0.50
16.	Iron (mgl^{-1})	0.01-0.30
17.	Hydrogen sulphide (mgl^{-1})	
	a. Fresh water	< 0.002
	b. Saline water	< 0.003
18.	Residual chlorine (mgl^{-1})	< 0.003

APPLICATION OF WATER FILTRATION TECHNIQUES

a) Air lift filter

The suspended impurities can be controlled by applying the air lift and rapid sand filters; while dissolved impurities can be removed by activated carbon pressure filters. Air lift filter is the most trouble free means of filtering water through synthetic sponge layer by pumping the water with airlift. In water holding aquaria, lift pipe extends below water level and filter chamber rests above the top water surface. The suspended impurities up to the size of 0.002 mm can be filtered out through this system. By pumping air @ 5 cm^3/sec., two litres of water/minute can be filtered when the diameter of the lift pipe is 1 cm. By doubling the diameter of the lift pipe, the rate of water flow can be enhanced by 5.6 times.

b) Mechanical filter/rapid sand pressure filter

It is the physical separation of concentration of suspended particulate matters from circulating water. It is accomplished by passing the water through suitable substrate that traps the particles. The trapped materials are then removed by various methods depending upon the type and design of the filter. Gravels reduce turbidity in water by trapping particulate matter and removing it from suspension. This is accomplished in two ways. First, suspended matter is physically trapped in the interstices among gravel grains. Second, the electrostatically charged surfaces of the gravel grains attract oppositely charged particles or colloids and remove them from water. Rough and angular gravels are best for mechanical filtration. Their numerous surfaces increase the electrostatic potential of the filter bed. The mechanical filtering efficiency of gravel increases with decreasing grain size of the individual granule. Smaller granules have more surface exposed to the water for electrostatic attraction of particulate matter. The smaller interstices facilitate removal of finer particles. These result in a greater percentage of suspended matter removed per unit volume of filtered water. Even distribution of each grade of gravel is very important when initially filling a sand filter. This assures well distributed circulation through the gravel column. No gravel bed should be lesser than 5 cm, no matter how small the culture system is. Systems lesser than 10,000 litres rarely need filter bed more than 50 cm thick if they are well designed and managed. These filters are applicable on large-scale water demand, *e.g.*, in carp hatcheries and in aqua farms, more particularly for slow sand mechanical filtration system. Rapid sand pressure filter system separates and concentrates particulate and dissolved organic carbon by trapping them against suitable filtering media of gravel, sand and activated carbon. The pressure filter is powered by mechanical pump instead of airlift and hence has faster clear water flow rate than airlift filter.

c) Activated carbon filter/chemical filtration

It is the removal of substances (primarily dissolved organics but also nitrogen and phosphorus compounds) from solution on a molecular level by absorption on a porous substrate, or by direct chemical fractionation or oxidation. The removal of dissolved organics by absorption is done by activated carbon or charcoal, which is a porous substance, and its degree of efficiency or absorptive capacity is measured by the total surface within the pores that are available to chemically attract organic molecules. In aquaculture, it is usually the granular type that is used rather than powdered activated charcoal. One Kg of a high grade powdered variety may contain several million square meter of surface area, but granular types are both chemically efficient and easier to handle. After the removal of colloidal and suspended solids, the soluble refractory organics may be efficiently removed by contact with activated carbon granules. Such carbon absorbs up to 20-30 percent of its own weight in mixed organics from wastewater. It is better to place glass wool on top of the carbon to reduce colloidal surface coating of the granules. Granular activated carbon is best kept in a separate container or contactor. Scattering it on the surface of a filter bed is impractical because once the material is saturated; it must be separated from the gravel grains and removed. Chemical filtration can make almost complete removal of dissolved impurities and they are applicable in prawn hatcheries, as they need very clean water.

Air lift and sand filters reduce the level of particulate matter. Biological filtration removes a portion of organics in culture water by mineralization but neither process is adequate to keep the level of dissolved organics within safe limits. It can only be accomplished by activated carbon filtration techniques. Granular activated carbon with a particle diameter greater than 0.1 mm is suitable for use. GAC is manufactured in two steps. The first is char formation in which a carbonaceous material is heated at about 600°C to drive off the hydrocarbons. Char formation is done in the virtual absence of air to prevent combustion. The second step is activation. The char is reheated this time to about 900°C in the presence of an oxidizing gas which develops the porous internal structure on which dissolved organic carbon is collected from water.

d) Biofilter/under gravel filter

Contrary to other devices of filtrations, bio-filter is a device in which filtration is effected by involving micro-organisms. It restores the quality of culture water by mineralization of organic nitrogenous compounds, nitrification and denitrification by bacteria suspended in the water and attached to the gravel in the filter-bed. Bio-filter should be kept away from sunlight to avoid algal growth, which may affect its working efficiency. Types of bio-filters used in aquaculture are trickling, submerged, updraft and rotating disc filters. Heterotrophic and autotrophic bacteria are the major groups present in aquaculture systems. Heterotrophic species utilize organic nitrogenous compounds excreted by the animals as energy sources and convert them into simple compounds, such as ammonia and nitrates.

The mineralization of these organics is the first stage in biological filtration. It is accomplished in two steps, ammonification, which is the chemical breakdown of proteins and nucleic acids producing amino acids and organic nitrogenous base; and deamination in which a portion of organic and some of the products of ammonification are converted to inorganic compounds. Once organics have been mineralized by heterotrophs, biological filtration shifts to the second stage, which is nitrification; it is the biological oxidation of ammonia to nitrite and nitrite to nitrate by autotrophic bacteria. These organisms, unlike heterotrophs, require an inorganic substrate as energy source and utilize carbon dioxide as their only source of carbon. *Nitrosomonas sp.* and *Nitrobacter sp.* are the principal nitrifying bacteria in aquaculture systems. *Nitrosomonas* oxidizes ammonia to nitrite whereas *Nitrobacter* oxidizes nitrite to nitrate. The third and last stage in biological filtration is denitrification. This process is a biological reduction of nitrate or nitrite to either nitrous oxide or free nitrogen. Denitrification can apparently be carried out by both heterotrophic and autotrophic bacteria. Bio-filter removes dissolved impurities up to a limited extent and they are applicable in intensive fish spawn and fry and prawn larval rearing systems.

e) Sequence of filter operation

The basic operation of pressure filter, dual media filter and activated carbon filter is as follows:

Mode of operation: All units operate in down flow mode, where water enters from the top, percolates through the media and treated water is collected from the bottom.

Service: The water to be filtered enters from the top, percolates downward through the filter media and is collected from the bottom.

Backwash: The water enters from the bottom of the filter chamber, passes through the media and is drained from the top. This is called as backwash and it is done to wash away the slimy dirt (schmutzdecke) which gets accumulated on the top of filter bed. The backwash is generally done once at every 24 hrs interval or when pressure drop exceeds 8 psi (0.562 kg/cm^2).

Rinsing: The water enters from the top passed through the filter media and is drained off from the bottom.

f) Rules for designing a filter

Calculate the area of vessel by required volumetric flow rate and the velocity as mentioned below.

Area (m^2) = Volumetric flow rate (m^3/hr)/Velocity (m/hr), based on above calculated area calculate dia. of the vessel by the following formula:

Dia. (m) = [Area m^2/0.7856]$^{1/2}$

Calculation: Q = KHA/L, m^3/sec.

Where:

Q = Discharge of the filter, m^3/sec.

K = Permeability coefficient, m/sec.

H = Total head, it is the difference in elevation between the surface of the source of water and the point of discharge, m.

A = Surface area of the filter, m^2.

L = Thickness of the filter medium, m.

Table 48: Permeability coefficient 'K' for different media

Sl. No.	Sand type	Average grain size (mm)	Range of 'K' (m/sec)
1.	Medium gravel	4 - 7	10^{-2} x 4 – 7 x 2 = 0.08 to 0.14
2.	Fine gravel	2 – 4	10^{-2} x 2 – 4 x 2 = 0.04 to 0.08
3.	Coarse sand	0.5 - 2	10^{-2} x 0.5 – 2 x 2 = 0.01 to 0.04
4.	Medium sand	0.3 – 0.5	10^{-2} x 0.3 – 0.5 x 2 = 0.006 to 0.01
5.	Fine sand	0.1 – 0.3	10^{-2} x 0.1 – 0.3 x 2 = 0.002 to 0.006

For sands, the permeability coefficient can be estimated from the Hazen's equation:

$$K = 10^{-2}D_{10}^{2}$$

D_{10} is the grain size in mm.

g) Important points on filtration system

1. Normally, pressure sand filter is used to filter out suspended solids up to 30 ppm and dual filter for 55 ppm and higher suspended solids require coagulation. Output quality of water from a pressure sand filter is 25 – 50 microns.

2. Velocity for water treatment in sand filter is taken as 7.5 – 18 $m^3/m^2/hr$ (residential filter) and 20 – 30 $m^3/m^2/hr$ in case of institutional filter. For recirculatory water pool, velocity can be taken more than 35 $m^3/m^2/hr$ for low turbidity application.

3. Higher velocity will induce higher head loss and in turn the frequency of backwash will increase.

4. Backwash of filter should be done using clean water.

5. Whenever air scouring is provided, it should be done before backwash.

6. If strainers are provided at the bottom, there is no need to put pebbles and gravels.

7. When the resistance in the filter due to clogging, *i.e.*, head loss is equal to the total depth of water on the filter, the operation will stop. The head loss should not be greater than depth of filter sand; when it becomes excessive and before a negative head is formed, the filter should be cleaned.

h) Ultraviolet radiations

Ultraviolet radiations kill microorganisms in water directly by deactivating DNA within the cells. UV lamps produce radiation and the process by which water is treated is called as irradiation. The effectiveness of a UV sterilizer depends on three factors:

1. Size of the organism, 2. Amount of radiation generated and 3. Penetration of UV rays in water. In general, larger is the organism, it is more resistant to UV radiations. Many viruses, bacteria and the smaller life stages of fungi and protozoans can be killed by irradiating them with 35000 $uWsec^{-1}cm^{-2}$. UV rays probably can not penetrate water farther than about 5 cm under ideal conditions. Penetrating power is further reduced by dissolved and particulate particles, and inorganic ions present in water. Thus, UV irradiation is less effective in waters that have higher concentration of TDS and TSS. Due to the presence of large concentration of inorganic ions in brackish water and sea water, UV irradiation is comparatively less effective in saline water than freshwater at similar radiation level. The installation of UV filter is done at the final or last stage of water treatment, *i.e.*, after the application of mechanical, chemical and biological steps. The last step, UV installation is required to get assured for problem free optimal water quality, particularly for prawn hatchery practices.

Table 49: Problems caused by adverse environmental conditions

Sl. No.	Causative agent	Symptoms	Preventive and remedial measures
1.	pH	i) Acidosis: fast swimming movements, gasping at surface, occasional jumps out of water, or extreme sluggishness, tendency to hide, loss of colour and appetite.	Avoid overstocking with fish and under stocking with plants. Carry out immediate partial water exchange. Add lime @ 5-10 ppm.
		Prone to fungal infections.	*Malachite green@ 0.001 ppm.
		ii) Alkalosis: serious damage to gills, disintegration of fin edges, general opaqueness of skin.	Relocate heavily planted tanks away from prolonged and direct sunlight. Carry out immediate partial water exchange. Apply alum @ 2-5 ppm.

[Table Contd.

Contd. Table]

Sl. No.	Causative agent	Symptoms	Preventive and remedial measures
		Prone to ectoparasitic infection.	30 minutes bath in 100 ppm formalin (1 ml in 10 litres).
2.	Chlorine	Restless movements, loss of balance.	Measure residual chlorine with the help of chloroscope and orthotolidine reagent. Dechlorinate aquarium water with sodium thiosulphate @7 ppm for the removal of each one ppm residual chlorine.
3.	Ammonia and nitrite nitrogen	Inflamed gills and fin edges, blood spots, loss of balance.	Make partial water exchange and apply aeration. Use ammonia absorbing medium such as zeolite (1ppm zeolite removes 1.5 ppm ammonia nitrogen, NH_3-N). Establish good biofilter system for nitrite reduction.
4.	Oxygen	i) Excess: Gas bubble disease, small bubbles visible under skin in fins and around head and eyes.	Regulate aeration. Relocate tank away from direct sunlight if overstocked with plants.
		ii) Insufficient oxygen, gasping at surface, partial loss of colour.	Initiate faster aeration. Carry out partial water exchange. Spray water on surface.

* Preparation of 0.001 ppm malachite green or 1 µg/l

Dissolve 1g of malachite green in 1 litre of distilled water (stock solution -1, it is 1 mg per 1 ml); take 1 ml of stock solution -1 and dilute up to 1 litre with distilled water (stock solution -2, it is 1 µg per 1 ml); 1 ml of stock solution -2 in a litre of aquarium water will be 0.001 ppm of malachite green dose.

Table 50: Effects of water quality parameters in the context of aquahatchery and production systems

Sl. No.	Parameter	Effect on fish productivity	Remedial measure
1.	pH	• Low pH increases the susceptibility of fish to disease. • Low pH decreases fecundity and eggs fertility. • Viability of fish spawn is largely dependent on pH. • Low pH water corrodes the gills and fins.	Application of lime for low pH and alum for high pH

[Table Contd.

Contd. Table]

Sl. No.	Parameter	Effect on fish productivity	Remedial measure
		• The acidic and alkaline death points for fish are 4 and 11, favourable range of pH is 7.5 – 9.0 (for freshwater); 8.0 - 8.5 (for saline water).	
2.	Suspended solids	• Chocking of gills, in turn the problem of asphyxiation thus reduces carrying capacity of oxyhaemoglobin.	Mechanical control and alum
3.	Total hardness	• Mortality in hatcheries of juveniles. • Eutrophication in culture system.	Zeolite, sodium carbonate, alum
4.	DO <1 mg/l	• Low D.O. accelerates the effects of toxicants, which enhance the disease probability.	Increase aeration
	1-5 mg/l	• Inadequate supply of oxygen hampers ovulation and in later stage results in death of embryos. • Lethal, if exposure lasts longer than a few hours. • Fish survive, but reproduction is poor and slow growth if exposure is continued.	
	> 5 mg/l	Fish reproduce and grow normally.	
5.	BOD/COD	BOD and COD cause reduction of DO solubility and availability, in turn enhance in total organic load.	Biofilter and aeration
6.	Ammonia nitrogen	• Toxic to fish. • Damage is confined to respiratory organs, blood and nervous tissues of fish. • Gills show hyperplasia. • Susceptibility to ectoparasitosis and myxobacterial infections.	Zeolite and dilution of water
7.	Total nitrogen	• Causes eutrophic conditions.	Alum, zeolite and dilution of water
8.	Phosphate	• Enhances organic load and eutrophication.	Alum, lime and zeolite
9.	Iron	• Precipitation of ferric hydroxide or ferric oxide on the gills of fish. • Precipitation of ferric hydroxide on incubating eggs may also smother and kill developing embryos. • Cause necrosis of the gills. • Makes the aqueous medium acidic.	Prechlorination @ 0.64 × iron content
10.	Viscosity	• Larval mortality in hatcheries	$NaNO_3$ @ 10 ppm

Table 51: Molecular and combining weights of principal compounds involved in water and wastewater treatment

Substance	Chemical formula	Molecular weight	Combining weight	Applicable to control
Aluminium hydroxide	$Al(OH)_3$	78	26	Turbidity, alkalinity and phosphate
Aluminium sulphate	$Al_2(SO_4)_3$	342	57	
Calcium bicarbonate	$Ca(HCO_3)_2$	162	81	Acidity, fluoride, phosphorus and pathogens
Calcium carbonate	$CaCO_3$	100	50	
Calcium hydroxide	$Ca(OH)_2$	74	37	
Calcium oxide	CaO	56	28	
Potassium permanganate	$KMnO_4$	158	31.6	Pathogens
Calcium sulphate	$CaSO_4$	136	68	pH- 4.0 -6.5 pH above 9.5 The zone of turbidity and oil
Ferric chloride	$FeCl_3$	162.5	54.2	
Ferric hydroxide	$Fe(OH)_3$	106.84	35.61	
Ferric oxide	Fe_2O_3	159.70	26.61	pH- 4.0 -9.5 pH 9.5 and above
Ferric sulphate	$Fe_2(SO_4)_3$	400	66.6	
Ferrous oxide	FeO	71.84	35.92	Iron and sulphate
Ferrous sulphate	$FeSO_4$	152	76	Heavy metals and acidity
Hydrochloric acid	HCl	36.5	36.5	
Magnesium bicarbonate	$Mg(HCO_3)_2$	146	73	
Magnesium carbonate	$MgCO_3$	84	42	
Magnesium hydroxide	$Mg(OH)_2$	58	29	
Magnesium sulphate	$MgSO_4$	120	60	
Sodium bicarbonate	$NaHCO_3$	84	84	Hardness, pathogens, acidity and oil
Sodium carbonate	Na_2CO_3	106	53	
Sodium chloride	$NaCl$	58.5	58.5	
Sodium hydroxide	$NaOH$	40	40	
Sodium silicate	Na_2SiO_3	122	61	
Sodium sulphate	Na_2SO_4	142	71	
Sulphuric acid	H_2SO_4	98	49	Iron, chlorides and alkalinity

N. B. When ferrous salts are to be used, addition of lime is necessary for hydrolysis of ferrous ions. Ferrous salts do react with natural alkalinity, but it is a delayed reaction. Therefore, lime is to be added first.

$FeSO_4 \cdot 7H_2O + Ca(OH)_2 = Fe(OH)_2 + CaSO_4 + 7H_2O$

Table 52: Atomic weight, valency and combining weight of elements and radicals involved in water and wastewater treatment

Element	Atomic symbol	Atomic weight	Valency	Combining weight
Aluminium	Al	26.98	+3	8.99
Barium	Ba	137.40	+2	68.70
Calcium	Ca	40.08	+2	20.04
Carbon	C	12.01	+4	3.00
Copper	Cu	63.54	+2	31.77
Hydrogen	H	1.008	+1	1.008
Iodine	I	126.91	-1	126.9
Iron(ferric)	Fe	55.85	+3	18.61
Iron(ferrous)	Fe	55.85	+2	27.92
Magnesium	Mg	24.32	+2	12.16
Manganic	Mn	54.93	+3	18.3
Manganous	Mn	54.93	+2	27.5
Nitrogen	N	14.01	Several	-
Oxygen	O	16.00	-2	8.00
Potassium	K	39.10	+1	39.10
Sodium	Na	23.00	+1	23.00
Sulphur	S	32.06	Several	-
Zinc	Zn	65.4	+2	32.7
Ion/radical		**Molecular weight**		
Ammonium	NH_4	18	+1	18
Bicarbonate	HCO_3	61	-1	61
Carbonate	CO_3	60	-2	30
Chloride	Cl	35.5	-1	35.5
Fluoride	F	19	-1	19
Hydroxyl	OH	17	-1	17
Nitrate	NO_3	62	-1	62
Nitrite	NO_2	46	-1	46
Phosphate	PO_4	95	-3	31.66
Silicate	SiO_2	60	-2	38
Sulphate	SO_4	96	-2	48
Sulphide	S	32.10	-2	16
Sulphite	SO_3	80.10	-2	40

Conclusion

Etiology includes the science of finding causes and origins of diseases; while posology comprises to determine quantification and administration of doses. Stress can lead to cause diseases and broadly these are of three types, 1. Preventive; 2. Curative and 3. Noncurative. Prophylactic treatment measures like liming, coagulation, sedimentation, aeration, chlorination, ozonation and ultraviolet irradiation are practiced to prevent the occurrence of diseases. The disease diagnosis requires having knowledge of clinical symptoms coupled with pathological investigations to decide the therapeutic measures to be taken for curative diseases. To get rid of from noncurative diseases, discard the infected ones out of system, as it is no longer useful to keep them in culture system; otherwise the disease problem will get further aggravated.

To cause disease trouble worse, poor water quality and inadequate nutrition are two very important factors; while healthy fishes have adequate resistance against pathogens and parasites. The nonconductive environment can lead to viral, bacterial, fungal, protozoan, worms and flukes infectious response. The common chemicals used for the recovery of fish diseases are sodium chloride, sodium nitrate, formalin, iodine, hydrogen peroxide, potassium permanganate, copper sulphate, EDTA, malachite green, methylene blue and organic poisons. To get best treatment results; sound knowledge on, chemical properties, their residual effects and mode of application, is necessary.

REFERENCES

Adelman, I. R. and Smith, L. L., 1972. Toxicity of hydrogen sulfide to gold fish, *Carassius auratus* as influenced by temperature, oxygen and bioassay techniques, NRC Research Press.

Alabaster, J. S. and Lloyd, R., 1980. Water quality criteria for fresh water fish. FAO publication. 297.

Alanson, B. R., Bok, A. and Van Wyk, N. I., 1971. The influence of exposure to low temperature on *Tilapia mossambica* Peters (Cichlidae). II. Changes in serum osmolarity, sodium and chloride ion concentrations. *J. Fish. Biol.,* 3: 181-185.

American Public Health Association, 2005. Standard methods for the examination of water and wastewater, 21st edition, Washington, DC, USA.

Banerjea, S. M., 1967. Water quality and soil condition of fish ponds in some states of India in relation to fish production. *Indian J. Fish.* 14(1&2): 115-144.

Beamish, R. J., 1972. Lethal pH for the white sucker, *Catostomus commersoni* (Lacepede). *Trans. Am. Fish Soc.,* 2: 355-358.

Boyd, C. E., 1974. Lime requirements of Alabama fish ponds. Alabama Agricultural Experiment Station Bulletin 459, Auburn University, Auburn, AL, 20 p.

Boyd, C. E., 1990. Water quality in ponds for aquaculture. Alabama Agricultural Experiment Station, Auburn University, Alabama. 482 p.

Boyd, C. E., 1995. Bottom soils, sediment and pond aquaculture. Chapman and Hall, New York. 348 p.

Boyd, C. E., 1998. Water quality for pond aquaculture. Alabama Agricultural Experiment Station, Auburn University. Alabama Research and Development Series No. 43:37 pp.

Brett, J. R., 1958. Implications and assessments of environmental stress. *In:* Investigations of fish-power problems. P.A. Larkin (ed.), H.R. MacMillan Lectures in Fisheries, University of British Columbia, 69-83 pp.

CAA (Coastal Aquaculture Authority), 1999. Guidelines for regulating coastal aquaculture. CAA, Chennai, 1-25 p.

CAA (Coastal Aquaculture Authority), 2002. Guidelines: effluent treatment system in shrimp farms. CAA, Chennai, 1- 22 p.

Central Inland Fisheries Research Institute, 1968. Methodology on reservoir fisheries investigations in India (CIFRI Bulletin no. 12), 102 p.

Central Institute of Brackishwater Aquaculture, 2016. Application of minerals in shrimp culture systems. CIBA Extension Series no. 52. ICAR-CIBA, 75, Santhome High Road, R. A. Puram, Chennai - 600028, India.

Central Institute of Fisheries Education, 1997. Recent advances in management of water quality parameters in aquaculture. Training Manual, 183 p.

Chandra Prakash, 2016. Productivity assessment, enhancement and criteria for fish seed stocking density in reservoir. Dept. of Fisheries, Govt. of Maharashtra, Mumbai, India. Training Manual, 55 p.

Chandra Prakash and Pawar Nilesh, 2010. Water quality management techniques in ponds and hatcheries. ICAR-CIFE, India, 111 p.

Chandra Prakash, Roy, S. Dam and Antony Jose, 2012. Water and wastewater treatment and utilization for aquaculture industry. ICAR-CIFE, India, 151p.

Chandra Prakash and Saharan Neelam, 2013. Jaliya samvardhan hetu pani avam mitti ki janch tatha prabandhan (in Hindi). ICAR-CIFE, India, 124 p.

Chandra Prakash, Sawant P. B. and Paul Lokesh, 2014. Unutilised and underutilised land and water resources for aquaculture industry. ICAR-CIFE, India, 174 p.

Chandra Prakash, Sinha, P.S.R.K. and Reddy, A. K., 1990. Economic viability of aquaculture in sewage. Journal of Environmental Biology 11 (1), 7-14.

Chandra Prakash, Verma, A. K. and Sawant, P. B., 2015. Utilization of degraded water resources through pisciculture. E-Book, Indian Agricultural Statistics Research Institute, New Delhi, India. 300 p.

Colman, J. A. and Edwards, P., 1987. Feeding pathways and environmental constraints in waste-fed aquaculture. Balance and optimization, *In*: D.J.W. Moriarty and Pullin, R. S. V. (ed.), Detritus and microbial ecology in aquaculture. ICLARM, Manila, 240-281.

Datta, M. K., Saha, R. D., Dhanze, R. J., Chandra Prakash, Kohli, M. P. S. and Saharan, N., 2008. Nutrient profile of pond water in Northeastern state of Tripura and impact of water acidity on aquaculture productivity, *Ind. Fish. Assoc.* 35:9-17.

Edwards, P., Kaewpaitoon, K., McCoy, E. W. and Chantachaeng, C., 1986. Pilot small scale crop/livestock/fish integrated farm. AIT Research Report, 184, AIT, Bangkok, Thailand.131.

Edwards, P., 1990. General discussion on wastewater-fed aquaculture. *In*: P. Edwards and R.S.V. Pullin (ed.), Wastewater-fed aquaculture, Proc. Int. Sem. Wastewater reclamation and reuse for aquaculture, Calcutta, India, December 6-9, 1988, ENSIC, AIT, Bangkok, 281-291.

Egna, H. S. and. Boyd, C. E., 1996. Dynamics of pond aquaculture. CRC Press. New York.

Einsele, W., 1956. Neue Erkenntnisse und wege dei der Erbrutung von Forelleneiern (Advances in hatching methods of trout eggs) *Osterr. Fishrii,* 9: 93-101.

EPA (Environmental Protection Agency), 1973. Water quality criteria, 1972. Washington, D.C., EPA-R-74-033 March 1973. 594 p.

FAO, 2000. FAO Year Book, Fishery Statistics, Agriculture Production, Vol. 86/2.

Fryer, J. L. and Pilcher, K. S., 1974. Effects of temperature on diseases of salmonid fishes. EPA Report 660/3-73-020, 112 p.

Haley, R., Davis, S. P. and Hyde, J. M., 1967. Environmental stress and *A. liquifaciens* in American thread fin shad mortalities. *Prog. Fish Cult.,* 29: 193.

Hendricks, B. and Jefferson, M. E., 1938. Structures of kaolin and talc-pyrophyllite hydrates and their bearing on water sorption of the clays: Arner. Min., V. 23, 863-876 pp.

Hicklings, C. F., 1962. Fish Culture, *Faber and Faber,* London, 225 p.

Jane, A. Russell, 1944. The colorimetric estimation of small amounts of ammonia by phenol: itypochlorite reaction. J. Biol. Chem. 156, 457-467.

Khanna, P. N., 2001. Indian Practical Civil Engineers' Handbook. Engineers' Publishers, Post Box 725, New Delhi – 110001.

Kumar, D., Raman, R. P. and Mohanty, A. N., 1996. Hydrogen sulfide mediated columnaris disease in Indian major carps in an indoor rearing system *In:* National seminar on Diseases in aquaculture, Kakinada, ICAR-CIFE, Mumbai (Abstract).

Lewis, W. M., 1960. Suitability of well water with high iron content for warm-water fish culture. *Prog. Fish Cult.,* 22(2): 79-80.

Lin, C. K., 1986. Acidification and reclamation of acid sulfate soil fish ponds in Thailand, *In:* Proc. of the First Asian Fisheries Forum, (Ed. Mclean, J., Dizon, L. B. and Hosilos, L. V.), *Asian Fisheries Society,* Manila, 71.

Llyod, R., 1961. Effect of dissolved oxygen concentrations on the toxicity of several poisons to rainbow trout (*Salmo gairdneri*). *J. Exp. Biol.,* 38: 445-447.

Lovell, T., 1973. Essentiality of vitamin C in diets of intensively cultured channel catfish. *J. Nutrition,* 103: 134-138.

Mackereth, F. J. H., Heron, J. and Talling, J. F., 1978. Water analysis: some revised methods for limnologists. Freshwater Biological Association, Scientific Publication No. 36. Titus Wilson & Son Ltd. Kendal 3000/3/78.

Mandal, L. N. and Chattopadhyay, G. N., 1992. Nutrient management in aquaculture. *In*: Non-Traditional Sectors for Fertilizer use (Ed. H. L. S. Tandon) FDCO, New Delhi, 1-17 pp.

Mathieson, A. McL. and Walker, G. F., 1954. Crystal structure of magnesium vermiculite. Amer. Min., v. 39, 231-265 pp.

McKee, J. E. and Wolf, H. W., 1963. Water quality criteria, Resource Agency of California, State Water Quality Control Board, Sacramento. Publ. No. 3-A, 548 p.

Menendez, R., 1976. Chronic of reduced pH on brook trout (*Salvelinus fontinalis*). *J. Fish Res. Bd. Can.,* 33: 118-123.

Milbrink, G. and Johansson, N., 1975. Effects of acidification of roe of roach, *Rutilus rutilus* L., and perch, *Perca fluviatilis* L., with special reference to the Avaa lake system in Eastern Sweden, Inst. Freshw. Res. Drottningholm, Report, 54: 52-62.

Miller, C. E., 2004. Soil Fertility. *Biotech Book*. New Delhi, India.

Ministry of Environment and Forests Gazette Notification, 1998. New Delhi, the 29th January, 1-13.

MrKenzie, R. M., 1970. The reaction of cobalt with manganese dioxide minerals. *Aust. J. Soil Res.* 8, 97-106.

Mukherjee, R. and Chattopadhyay, G. N., 2002. Efficiency of soil specific pond fertilization on fish production, *In*: The Fifth Indian Fisheries Forum proceedings, (Ed. Ayyappan, S.) AFSLB, Mangalore: 5-7.

Neess, J. C., 1949. Development and status of pond fertilization in Central Europe. *Trans. Am. Fish Soc.,* 76: 335-358.

Nishioka, R. S., Grau, E. G. and Bern, H. A., 1985. In vitro release of growth hormone from the pituitary gland of tilapia, *Oreochromis mossambicus. Gen. Comp. Endocrinol.* 60, 90-94.

Panda, N., Prasad, R. N., Mukhopadhyay, A. K. and Sarkar, A. K., 1988. Management of acid soils, 53rd Ann. Conf. Indian Soc. Soil. Sci. Bhubaneswar, India.

Patwardhan, A. D., 2013. Industrial wastewater treatment. PHI Learning Private Limited, New Delhi – 110001.

Prasad, R. N., Patiram and Munna, R., 1985. *Journal of Research, Assam Agricultural University.* 3:131.

Pullin, R. S. V., Rosenthal, H. and Maclean, J. L., 1992. Environment and aquaculture in the developing countries. Summary report of the Bellagio Conf. Environment and aquaculture in the developing countries, ICLARM. Conf. Proc. ICLARM, Manila, 16 pp.

Reddy, A. K., Chandra Prakash and Sinha, P. S. R. K., 1991. Larval rearing of *Macrobrachium rosenbergii* (De Man) in artificial sea water. Journal of Indian Fisheries Association, 21: 1-4.

Reddy, A. K., Chandra Prakash and Uniyal, R. P., 2000. Giant freshwater prawn culture. ICAR-CIFE, India, 217 p.

Romaire, R. P., Boyd, C. E. and Collis, W. J., 1978. Predicting night time dissolved oxygen decline in ponds used for tilapia culture. *Trana. Am. Fish. Soc.,* 107: 804-808.

Saharan, N., Chandra Prakash and Raizada Sudhir, 2002. Soil, water and aquatic biota analysis – a methods manual. CIFE, India, 128 p.

Schaperclaus, W., 1991. Fish Diseases, Vol. 12, Oxonian Press Pvt. Ltd., New Delhi and Calcutta, XVI, 597-1398 p.

Sehgal, J. L., 1993. Red and laterite soils of India: An overview. *In*. Red and laterite soils of India. NBSS and LUP. Nagpur, India.

Selye, H., 1936. A syndrome produced by diverse nocuous agents. *Nature,* 138: 32.

Selye, H., 1946. The general adaptation syndrome and the diseases of adapdation. *J. Clin. Endocr.,* 6: 117-230.

Selye, H., 1950. Stress and general adaptation syndrome. *Brit. Med. J.,* 1: 1383-1392.

Selye, H., 1956. The stress of life. Mc Graw-Hill Book Co., Inc. New York.

Selye, H., 1973. The evolution of the stress concept. *Am. Sci.,* 61: 692-699.

Seymour, A. H. and Donaldson, J. R., 1953. Occurrence of high mortality among chinook salmon fry after the recharge of charcoal filter. *Prog. Fish Culturist,* 15: 121-125.

Sinha, V. R. P. and Ramachandran, V., 1985. Fresh water fish culture. Indian Council of Agricultural Research, New Delhi.

Spotte, S. H., 1970. Fish and invertebrate culture – water management in closed systems. Wiley Interscience, John Wiley & Sons Ltd.

Strickland, J. D, H. and Parsons, T. R., 1972. A practical handbook of seawater analysis. 2nd edition. Bull. Fish. Res. Bd. Can.

Swan Environmental Pvt. Ltd., 2005. Swan diary. Hyderabad, India.

Train, R. E., 1979. Quality criteria for water. Castle House Publications Ltd. Billing & Sons Ltd., Guildford, London & Worcester.

Walkley, A. and Black, C. A., 1934. Estimation of soil organic carbon by the chromic acid liberation method. Soil Sci. 37: 29 – 38.

Wedemeyer, G., 1970. The role of stress in the disease resistance of fishes and shellfishes, Spec. Publ. No.5. *Am. Fish. Soc.,* Washington, D.C., 30-35 pp.

Wedemeyer, G. A. and Yasutake, W. T., 1978. Prevention and treatment of nitrite toxicity in Juvenile steelhead trout (*Salmo gairdneri*). *J. Fish Res. Board Can.* 35: 822-827.

Wedemeyer, G. A., Meyer, F. P. and Smith, L., 1999. Environmental stress and fish diseases. Narendra Publishing House, Delhi, 192 p.

Welch, P. S., 1948. Limnological Methods. Mc Graw-Hill Book Co. New York.

Welch, P. S., 1952. Limnology Methods. Mc Graw-Hill Book Co. New York. 538 p.

Wheaton, F. W., 1977. Ponds, tanks and other impounding structures. *In:* Aquaculture Engineering, John Wiley and Sons, New York, 459-483 pp.

Whitfield, M., 1971. Ion selective electrodes for the analysis of natural waters. Sydney. Australian Marine Sciences Association.

WHO, International Programme of Chemical Safety, 1989. Guidelines for drinking water quality. Recommendations. 2nd Ed. Vol. 1, World Health Organization, Geneva.

Winkler, L. W. C., 1888. The determination of dissolved oxygen in water. Berlin Deutsch Chem. Gesellsch 21: 28-43.

Yavalkar, S. P., Bhole, A. G., Vijay, B. P. V. and Chandra Prakash, 2013. Optimization of lime-soda process parameters for reduction of hardness in aquahatchery practices using Taguchi Methods. J. Environ. Sci. & Engg. Vol. 54(2): 260-267.

''कृण्वन्तो वारिणी मत्स्यम''

"MAKE ALL THE WATER BODIES FISH INHABITABLE"

"ANY PROBLEM IS TIME AND SOLUTION IS HUMAN"

'Chandra Prakash'